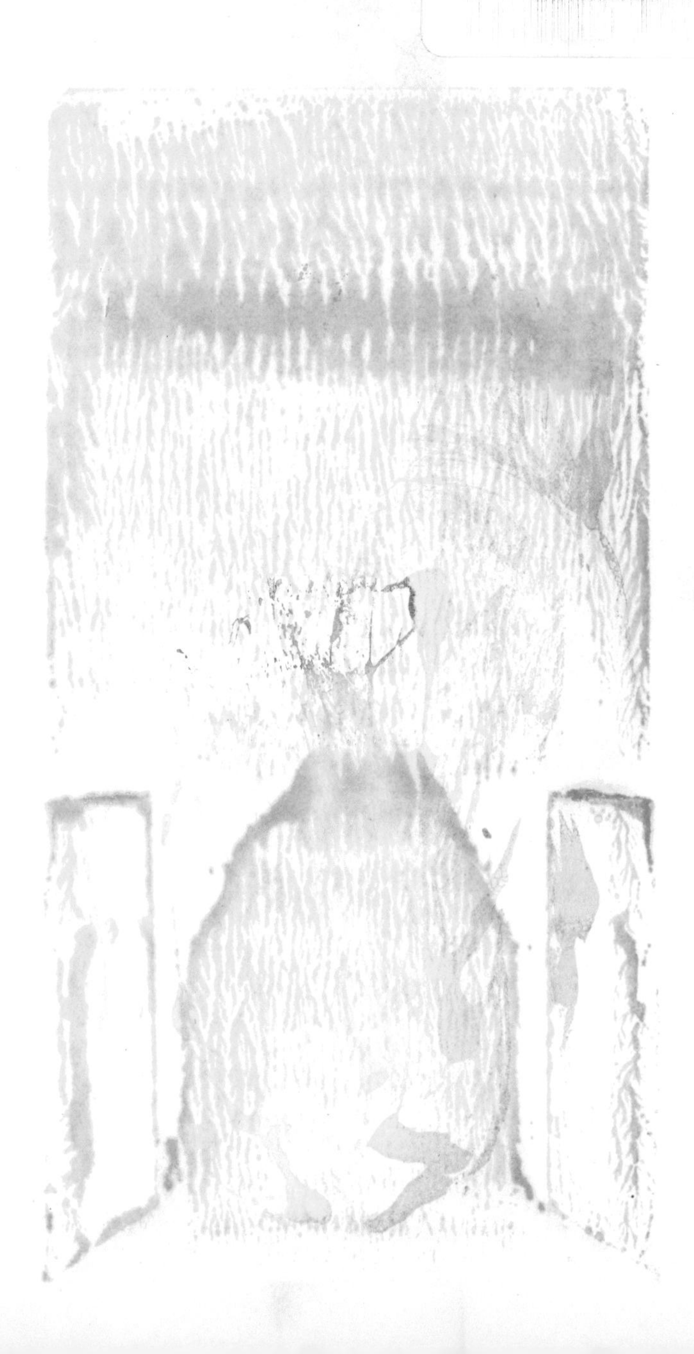

Fossils and the Life of the Past

Erich Thenius

Translated from the German

by

Barbara M. Crook

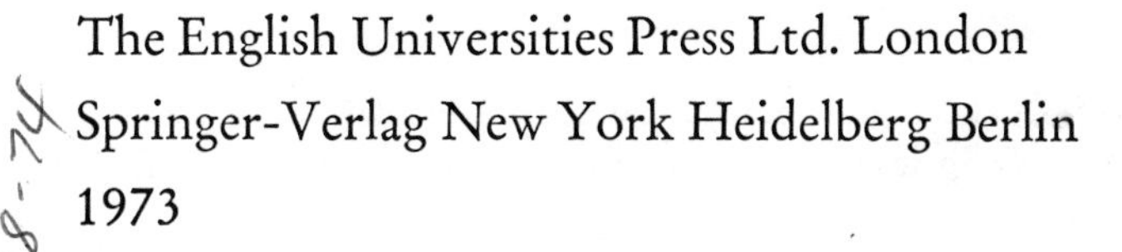

The English Universities Press Ltd. London
Springer-Verlag New York Heidelberg Berlin
1973

Erich Thenius, Ph.D.
Professor of Paleontology
Institute of Paleontology
University of Vienna
Vienna, Austria

Library of Congress Catalog Card Number 76-183484.

Printed in the United States of America.

ISBN 0-340-16970-2 The English Universities Press Ltd. London

ISBN 0-387-90039-X Springer-Verlag New York · Heidelberg · Berlin
ISBN 3-540-90039-X Springer-Verlag Berlin · Heidelberg · New York

To the memory of

O. Abel

and

E. Dacqué

Preface

This book first appeared in German under the title *Versteinerte Urkunden — Die Paläontologie als Wissenschaft vom Leben in der Vorzeit* in Springer-Verlag's series, *Verständliche Wissenschaft,* in 1963. The text has been slightly expanded and brought up to date for the English edition.

Prof. D. E. Dacqué's *Das Fossile Lebewesen — Eine Einführung in die Versteinerungskunde,* which was published in the same series in 1928, had been out of print for years. Moreover, there have been important advances in paleontology and related sciences in the last several decades; thus, a new popular presentation of this fascinating field of knowledge seemed necessary.

This volume is not a systematic, chronologically arranged survey of paleontological records; it is, rather, an attempt to give the general reader some idea of the importance, the working methods and the aims, not to mention the limitations, of paleontological research. If I succeed in making it clear to the layman that paleontology is no dry museum science, I shall have achieved what I set out to do. For paleontology, as will be shown in this book, not only plays a very important role in practical geology, through micropaleontology and palynology, but it is also of even more vital importance to the life sciences in helping to settle fundamental questions concerning evolution.

I am greatly indebted to many of my colleagues and to a number of institutions for help in compiling this volume and especially in assembling the illustrations. Among them are Prof. B. Accordi, Rome; A. Bachmann, Vienna; Prof. F. Bachmayer, Vienna; Prof. Z. Burian, Prague; Dr. E. H. Colbert, New York; Prof. E. Flügel, Darmstadt; Prof. A. Kieslinger, Vienna; Prof. W. Klaus, Vienna; Dr. G. Krumbiegel, Halle a. d. Saale; Prof. E. Kuhn-Schnyder, Zürich; Dr. F. Rögl, Vienna; Prof. A. Seilacher, Tübingen; † Prof. E. J. Slijper, Amsterdam; Dr. K. Staesche, Stuttgart; † Prof. R. A. Stirton, Berkeley; Dr. H. Stradner, Vienna; Dr. W. Struve, Frankfurt a. M.; Prof. W. Stürmer, Erlangen; Prof. E. Voigt, Hamburg; Prof. H. Zapfe, Vienna; Dr. A. Zeiss, Erlangen; Musée National d'Histoire naturelle de Paris; Museum of Paleontology, University of California, Berkeley; Naturmuseum und Forschungsinstitut Senckenberg, Frankfurt a. M.; the American Museum of Natural History, New York; Staatliches Museum für Naturkunde, Stuttgart; Petrified Forest National Monument, U.S. National Park Service; United States Information Service, Vienna.

I thank all concerned for their willingness to help me. For assistance in reading the proofs, my thanks are extended to Miss M. Tschuggüel, and for the preparation of the photographs to Mr. F. Sattler.

I also thank Springer-Verlag for its care in designing the book and preparing the drawings and Dr. D. A. B. Pearson, Laurentian University of Sudbury, for editing the English text.

Vienna, *June 1971* ERICH THENIUS

Contents

Fossils and the Life of the Past

I. Introduction

Definition of Paleontology. Its Significance

Taken literally from the Greek (palaios — old, on — being, logos — science), paleontology is the science of life long ago. This, however, is by no means an adequate description of the sphere of activity of paleontology. The science of life in the geological past, it is not to be confused with either *archeology,* the study of ancient cultures, or *prehistory.* For whereas archeology and prehistory are concerned exclusively with the traces left by man and may be numbered among the humanities, *paleontology* is one of the *natural sciences,* and its function is to study the prehistoric plant and animal world. The limits of paleontology, as compared with those of prehistory and archeology, are determined first and foremost by its subject matter and also to some extent by its time scale, as a glance at the Geological Time Scale will show. Any parallels that exist are to be found in the circumstance that all three disciplines are seeking to reconstruct a complete picture from what are usually only fragmentary remains.

As Fig. 1 shows, paleontology is complementary to zoology and botany; it is therefore subdivided into paleozoology and paleobotany. Another subdivision is micropaleontology which might just as well be called applied paleontology, in the same way that economic geology is applied geology. The division is not justified because of methods and content, but because of its special working techniques. While the limits that apply to zoology and botany may seem equally artificial, being purely temporal, in practical terms paleontology requires completely different methods of investigation. It is these methods which make it necessary to consider paleontology a separate discipline. Here then, we have a slightly different relationship from that between prehistory and archeology, where the distinction is based upon the absence or presence of written records of a culture, although their methods and objectives are essentially the same.

The zoologist and the botanist can make direct observations of animals and plants, their life forms, and their behavior and may even carry out experiments with living organisms. All this is denied the paleontologist who in his studies of the records of the past must assemble what knowledge of geological times he can find and use it to create a picture of plant and animal life as it used to be; thus, paleontology and geology together form the historic foundation of the natu-

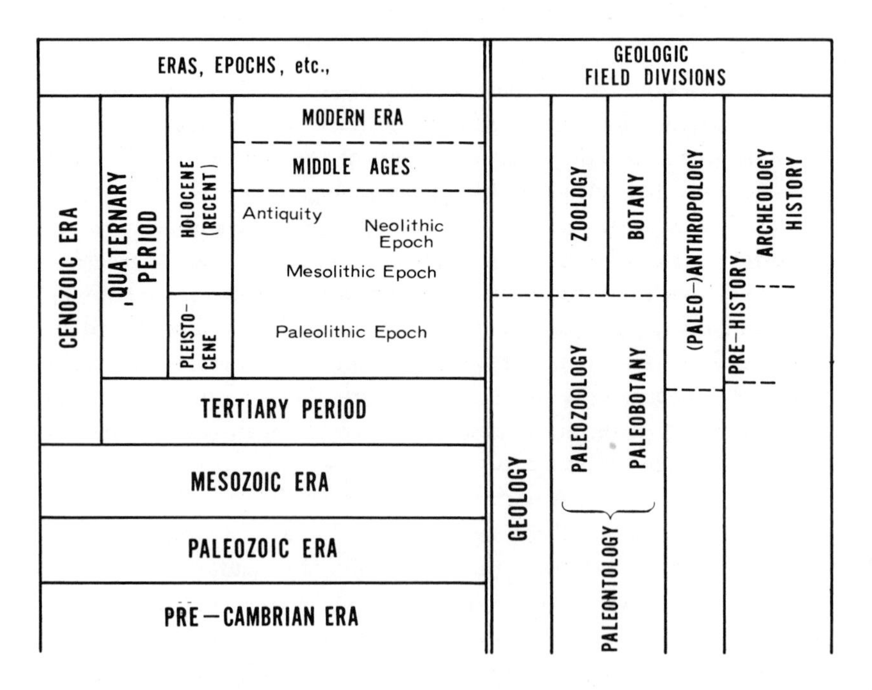

Fig. 1. Diagram showing the respective spheres of interest of paleontology, botany, zoology, and archeology.

ral sciences. This fact explains the close link between paleontology and geology, although paleontology, as a biological discipline, has a special significance for the natural sciences because of the light it sheds on the evolution of species.

So far we have outlined the essential tasks and aims of paleontology, although we shall go into them in more detail. But it is not enough merely to record and describe forms of life, to know where they were found, where they were distributed, and how they lived: paleontology also embraces a knowledge of the time scale of their existence and of the phylogenetic links between them, i. e., the history of the various species. Indeed, we may say that research into the origin of species falls within the special province of paleontology. It is thus an indispensable link in the chain of the life sciences. Only the privilege of looking backward can make biology a whole science; only paleontology can supply the evidence of the evolution of the animal and vegetable kingdoms.

For geology, too, paleontology performs valuable services — it enables the deposition of sedimentary rocks to be dated by means of index fossils; it is, in fact, the indispensable key to unlocking all the sedimentary treasures of the Earth.

Paleontological Sources

What is the *paleontological record?* At one time it was known as *petrifaction,* but this is not really the appropriate label, nor is the term "science of petrifaction" quite fair to describe the conception and scope of paleontology. The paleontologist prefers to speak of the objects he studies as *fossils,* but without thereby implying anything about their state of preservation, because this can vary tremendously and has often been the cause of faulty deductions.

So what then are fossils? All remnants of prehistoric organisms, including the traces of their activities (called trace fossils: tracks, burrows, etc.) are called fossils, a distinction being made between fossil bodies and fossil traces. By "prehistoric" we mean all remains of plants and animals that lived before the geological present, or Holocene (see Geological Time Scale, p. 180). Thus, fossils can be remains from the Ice Age or even from Precambrian times.

In this sense, "fossil" does not refer to the state of preservation e. g., a greater or lesser degree of mineralization, or, as it is often incorrectly called, to the degree of fossilization (incidentally, the petrified leaves, etc. which are formed at the present time in calcareous springs are *not* fossils); nor does "fossil" imply that the species represented is extinct. The force of the term is simply that of a limitation in time, and it may be regarded as the opposite of the description Recent as applied to Holocene organisms. This also takes into account the original meaning of the word fossil, derived from the Latin verb, *fodere,* to dig (up). To be sure, the word was first used in a different sense. For the German doctor and mining engineer, Georg Bauer (1494–1555; his name was latinized to Agricola) the "father of mineralogy," who coined the word, took it to mean not only "petrifacts" of plants and animals but also stones, minerals, and artifacts (objects made by man), and hence any object of interest to geologists, mineralogists, and prehistorians. A. G. Werner, too, in his book *On the External Characteristics of Fossils,* published in 1774, took the word "fossil" to refer primarily to stones. Today, however, it is taken to mean the remains of prehistoric plants and animals.

As we shall show in the next chapter, many fossils cannot be described simply as petrifacts, since either the soft tissues are preserved, or only the impression of organic remains survives. True, by far the greatest number are hard tissues which have been preserved as fossils. What records are to the historian, fossils are to the paleontologist.

Historical Changes in Ideas about Fossils

Fossils were not always correctly identified as the remains of prehistoric forms of life, while, conversely, inorganically formed aggregates were often taken to be fossils. In antiquity, fossils were correctly regarded by some naturalists or historians (e.g., Xenophanes, Herodotus, Xanthos, and Strabo) as the remains of earlier forms of life and interpreted accordingly.[1] Unfortunately, this idea was by no means universal among the scholars of that time. So it was mainly due to the influence of the writings of Aristotle (384–322 B. C.), the best-known naturalist of antiquity, that throughout the Middle Ages and up until relatively recent times no progress was made in knowledge about fossils. People were content to pass on the opinions of Aristotle, while devoting their ingenuity to Scholastic expositions. Even in the sixteenth century, fossils were regarded, following Aristotle, as oddly

---

[1] Finds of fossil snails, used for personal adornment, from various palaeolithic sites in France, Germany, Czechoslovakia, and Austria, bear witness to the fact that the men of the late Stone Age recognized fossils and collected them. We cannot tell, of course, whether Stone Age man knew they were the remains of earlier forms of life, and it is perfectly possible that he may have used the shells of contemporary species of snails for the same purpose.

shaped stones, sports of nature *(lusus naturae)*, created spontaneously by a plastic force *(vis plastica)* present in the primeval mud. Their true nature was recognized by only a few philosophers and scholars. Among them was Leonardo da Vinci (1452–1519) who, as a builder of canals, correctly interpreted the presence of marine fossils in the Po Valley as a sign that this region was once covered by the sea. Then there were G. Fracastoro (1483–1553) and B. Palissy (1510–1590) who realized that the shells of clams and snails, as well as fish bones, could be preserved by mineralization.

Since this was the state of affairs, the theory put forward by the Diluvialists, as the believers in the Deluge theory (from the Latin *diluvium*, flood) were called, was a distinct advance upon the ideas of the Schoolmen. The Diluvialists, who were well represented about the end of the seventeeth and beginning of the eighteenth centuries, appreciated that fossils were survivals of earlier forms of life, but claimed that these had perished in the biblical flood and had been preserved as "witnesses" of the truth of the account of creation in the Bible. The most successful proponent of the Diluvialist theory was the Swiss physician and naturalist J. J. Scheuchzer (1672–1753), who in 1726 claimed to have discovered in the Miocene deposits at Oeningen on Lake Constance "the skeleton of a Child of Damnation, for the sake of whose sins disaster engulfed the whole World." It was not until about a century later that G. Cuvier identified this "*homo diluvii tristis testis*" as the fossilized skeleton of a giant salamander.

The origins of Diluvianism can be traced as far back as the middle of the sixteenth century. About the time it was in fashion, Nicolaus Steno (1638–1686) became the first scholar to have a clear idea of the importance of sediments for the history the Earth and to look upon fossils as the remains of living creatures. He formulated the law of stratification which states that, where there are undisturbed strata in a profile, the lowest stratum must be the earliest and the highest the most recent.

It was not until the eighteenth century, however, that the correct interpretation became generally accepted. In that century, and in the following one, the collecting of petrifacts became fashionable and so the study of petrifacts, as paleontology was then called, was born. The main object of the study of petrifacts was to describe and classify fossils; in the process of classification scholars became aware that long ago very many widely differentiated types of fauna had existed and that each era had its characteristic petrifacts. This view was first propounded by the English engineer William Smith (1769–1839), the founder of stratigraphy; he also initiated its practical application for age determination. About this time, too, Leopold von Buch (1774 to 1832) introduced the concept of index fossils. Most collectors of petrifacts were either students of geognosy, as geology was then called, or were engaged in mining, and so the fossils were of practical use to them as time indicators (cf. Chapter V).

Also about this time, the great Swedish naturalist Carl von Linné, or Linnaeus as he is better known (1707–1778), introduced his binomial nomenclature (see p. 51) for the Recent plant and animal species known at that time. This formed the basis of international communication and was later extended to take in fossils as well. So now the stage was set for systematic studies. But it was not until the French zoologist G. Cuvier (1769–1832) made his fundamental discoveries of fossil vertebrates that the true foundations of scientific paleontology were laid. The actual word "paleontology" was first used in the first half of the nineteenth century by Fischer von Waldheim and since then has won general acceptance. Cuvier, with his studies of the comparative anatomy of Recent and fossil vertebrates, established that in the course of the history of the Earth a succession of widely differing fauna had existed, that an increasingly complex level of organization could be recognized in them, and that there were indeed truly extinct species. [2] For instance, Cuvier showed that the Pleistocene mammoth was specifically different, i. e., of a different species, from the two living species of elephant. Furthermore, Cuvier's studies in comparative anatomy led him to formulate his so-called correlation law, which states that all parts of an animal must bear a definite relationship to one another, hence that conclusions concerning the total structure of the individual could be drawn from isolated bones. He published the results of his comprehensive investigations in a work which has now become a classic, entitled *Recherches sur les ossemens fossiles* (first edition 1812). Cuvier is rightly known as the father of vertebrate paleontology.

Unlike his rivals, J. B. de Lamarck and Etienne Geoffroy St. Hilaire, Cuvier persisted in his belief in the immutability of species.

---

[2] Diluvialists had maintained that there were no extinct species. They insisted that it was necessary to reconcile fossil species with living species.

He explained the diversity of fossil faunas as the result of local catastrophes. He thus found no transitions between the early Tertiary vertebrates as they follow each other in time in the deposits of the Paris Basin. This view was eventually elaborated by the paleontologist Alcide d'Orbigny (1802–1857) into a Catastrophe Theory (Catastrophism), which postulated numerous global catastrophes, each one of which exinguished all life, so that each time creation had to begin anew.

It was not, however, until the theory of evolution, or of the Origin of Species, which is indissolubly linked with the name of Charles Darwin (1809–1882), that the breakthrough occurred with the idea of the variability of species and the step-by-step evolution of life from one form to another.

Before this, though, the young G. L. Leclerc de Buffon in the second half of the eighteenth century had speculatively advanced the idea that all animals were created according to a uniform scheme of organization; Lamarck (1744–1829) and to some extent St. Hilaire (1772–1844) had put forth the view that there could be a real transformation in the forms of organisms, but they failed to produce any evidence.

In Darwin's time the contribution paleontology could make to the story of evolution was underestimated, and justifiably so, because the fossil material available was so limited. Since then, however, new fossil finds have not only confirmed the fundamental soundness of the theory of evolution but have at the same time made paleontology the most reliable prop of phylogeny. This reversal is on the one hand due to the intensive and systematic research carried out over entire regions of the Earth — North America, North, East and South Africa, Central and East Asia, South America, and Australia — and on the other is not unconnected with the twentieth century's heightened interest in drilling for oil, which has produced an enormous quantity of microfossil-containing cores. Thus, paleontology in the last decade presents quite a different picture from that of only 50 years ago.

This brief historical review also makes it clear that paleontology has sprung from two sources: geological and biological. This fact is still reflected today in the departmental organization of universities, where work of a predominantly stratigraphic nature is undertaken by the geology department, while the more biologically oriented vertebrate paleontology, and paleobotany, too, is the province of the departments of zoology and botany. This is true of many universities and institutes, for example, Harvard University in Cambridge, Massachusetts, and the Senckenberg-Forschungs-Institut in Frankfurt-am-Main, German Federal Republic.

Here, then, is yet another proof that the field of paleontology lies midway between that of geology and that of biology. The following chapters will make it clear that fossils can be adequately interpreted only in the light of both these related disciplines.

Paleontology thus emerges as a discipline in its own right within the natural sciences, a discipline whose purpose is to build up a body knowledge about the world of fossil plants and animals. Moreover, it is also an indispensable auxiliary science for various faculties to which it is marginal: it aids the prehistorian by identifying the fauna (including the trace fossils, see p. 116) and flora of paleolithic times; it aids the paleoanthropologist by dating human fossil finds; it aids the geographer by producing fossil proofs from the animal and vegetable kingdoms which define paleoclimate and paleogeography; it aids the (stratigraphic) geologist by providing index fossils; and finally, it aids the zoologist and the botanist by presenting evidence for the evolution of species.

Chapters IV–IX will offer selected examples in an attempt to demonstrate the power of interpretation of paleontological records when properly analyzed and to define the limits of scientific evaluability.

II. Fossilization. State of Preservation and Sites of Fossil Remains

Conditions for the Occurrence of Fossils

Before discussing the analysis and interpretation of fossils, we must say something about the genesis of fossils, their state of preservation, and the sites where they are found. If we ask: What is capable of preservation as a fossil?, the paleontologist can only answer: Given the right circumstances, absolutely anything. To be sure, the conditions for the preservation of soft tissues and other perishable remains are very rare indeed. As a rule, it is the hard parts of organisms which survive, such as the shells of snails and bivalves and the hard outer casing of crabs, trilobites, and other arthropods; the skeletal remains of sea urchins and other echinoderms; and the bones and teeth of vertebrates.

As for the conditions which permit fossilization, current experience teaches us that on land the bodies of dead animals are usually quickly destroyed by mechanical or biochemical processes. The soft tissues either decay, putrefy, or are eaten by scavengers, so that usually only the hard parts are left. These, too, as with a vertebrate animal whose skeleton consists of a large number of separate bones, become scattered with the passage of time and are destroyed. Of the countless wild animals that die in the course of a year, very, very few are found as dead bodies or even as bones. For the process of decay to be prevented or slowed down, some form of burial is necessary. The sooner after its death an organism is covered by sediment, the better are the chances of its becoming fossilized. Consequently, fossils are most frequently found where rapid sedimentation predominates, that is, in shallow seas, lakes, ponds, swamps, and bogs; in river and cave deposits, fissure fillings, and airborne sediments (e. g., loess); and in permafrost.

True, burial does not completely arrest the processes of decomposition, for when air is excluded decay continues anaerobically, destroying the soft organic tissue. Burial does not even protect the hard parts from eventual destruction. One can therefore make a distinction according to the time of destruction: *before, during*, or *after* fossilization. As may be seen from Fig. 2, the events during fossilization result in quite different states of preservation, some of these being highly characteristic. The principal states of preservation are bodily preservation, casts, impressions, and pseudomorphs.

Fossil Diagenesis and State of Preservation

Fossilization where the *hard parts* — teeth, bones, shells, etc. — have been bodily preserved to any extent basically involves a substitution of inorganic material for the original organic substance; this substitution may be associated with recrystallization of the mineral substances making up the skeleton [1] as, for example, occurs in the fossil skeletal remains of echinoderms to produce characteristic

[1] This most commonly involves the conversion of the unstable form of calcium carbonate into the stable form (aragonite → calcite).

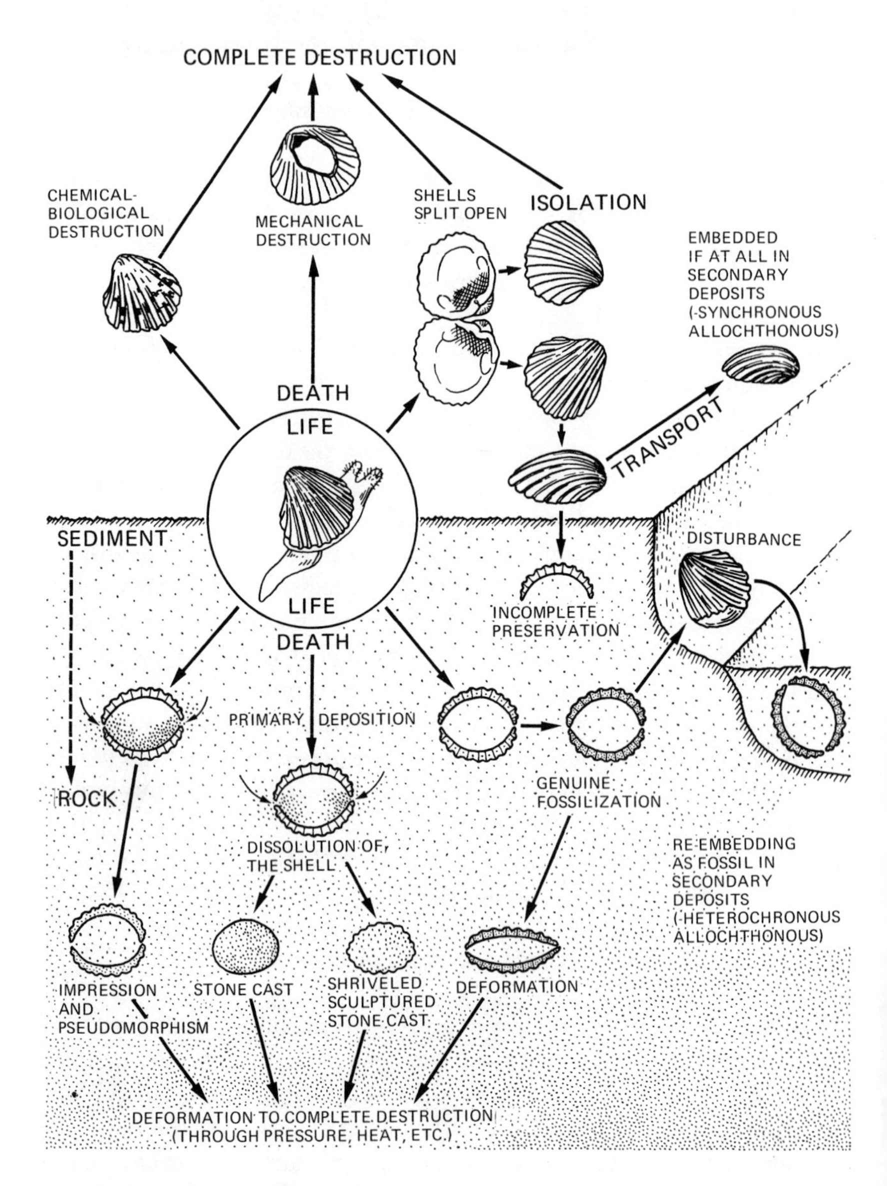

Fig. 2. Diagram showing the process of fossilization. Fossilization requires embedding in sediment. *Conditions of preservation:* impression, core (Steinkern), true petrification, natural cast. Note autochthonous (primary) and allochthonous (secondary) deposits. Synchronous allochthony by transport *before* fossilization; heterochronous allochthony by transport *after* fossilization.

rhombohedral fracture planes. The main inorganic components involved in the substitution are carbonates and minerals of the elements silver (Ag), copper (Cu), lead (Pb), zinc (Zn), iron (Fe) and sulfur (S), and silica, or silicic acid (SiO_2). In the fossilization of plants, the following are also involved: humic acids for lignification (part of the process of the formation of coal and bitumens), oil, asphalt, and native paraffin. With animal fossils, the most important agents are carbonates: calcium carbonates ($CaCO_3$) mainly in the form of calcite and less commonly as aragonite, dolomite ($MgCaC_2O_6$), and silica. A minor occurrence in fossil diagenesis is the conversion into pyrite or marcasite (FeS_2), limonitization by iron hydroxide which may be in the form of brown ironstone, and conversion to sulfates (gypsum $CaSO_4 \cdot 2\,H_2O$), phosphates [vivianite $Fe(PO_4)_2 \cdot 8\,H_2O$], and sometimes glauconite. Interestingly enough, there are still traces of organic substances, even in Paleozoic fossils, in the form of the amino acids, glycine, alanine, glutaminic acid, and aspartic acid. This is also true of certain organic dyes which seem to be even longer-lived, since they have actually been found in Precambrian shales. This is the field of *paleobiochemistry.* Although it is only now being developed, paleobiochemistry has been able to show that the basic organic materials were already present in Precambrian times. The preservation of organic substances in fossils is most clearly demonstrable in bituminous deposits. Thus, porphyrins from bituminous layers of the Triassic of Monte San Giorgio are as well attested as, say, chlorophyll from the Geiseltal Eocene.

In the course of fossil diagenesis it may happen that not only are the organic substances replaced by inorganic, but both may subsequently disappear, particularly where readily soluble components of the hard parts are concerned. This phenomenon, known as *leaching diagenesis,* is found in a variety of deposits and results in deformation without fracture. Usually this involves decalcification, which makes the initially rigid hard parts flexible. In such cases, if at least some portion of the original structure is retained, we call it pseudomorphism,[2] a complete substitution of the original remains.

[2] Many paleontologists understand by "pseudomorphism" fossils in which substantial alteration has occurred, with or without structural changes (e.g., alteration of aragonite to calcite). To some extent, stone casts may also be regarded as pseudomorphs. Occasionally, too, the shells of epibionts (e.g., clams—*Dimyodon, Pycnodonta)* which imitate the surface of the host animal may be called pseudomorphs.

The formation of pseudomorphs, which is shown schematically in Fig. 2, can be traced as follows: the remains embedded in the sediment, subsequently turned into rock, are completely dissolved by infiltrating water, and the space so created is secondarily filled with crystals. Examples of artificial pseudomorphs, or casts (Steinkern), are the bodies of the inhabitants of Pompeii who were buried by volcanic ash when Vesuvius erupted in 79 A. D. It is not the actual bodies that are preserved in the volcanic tuffs, which have hardened in the meantime, but merely the cavities representing the space they originally occupied; when the site was excavated, plaster casts were made from these cavities so as to give some idea of what the Pompeians looked like. Much more frequent than true pseudomorphs are cores or so-called steinkerne. These are fillings of hollow spaces which were usually, though not always, occupied by the soft tissues of the living organism (Fig. 3). The appearance and completeness of cores depends very much upon the properties of the filling material, which is generally sediment and more rarely crystallized mineral salts. The finer the

Fig. 3. Shell and core of an oyster from the Burdigalian at Retz (Lower Austria); size reduced. Orig. Paleont. Inst., Vienna.

particles of the filling material, the more faithfully the details are preserved. Casts that reproduce the external details of a shell, which has subsequently been dissolved away, are called replica. Some authors use this term to mean thin-shelled fossils whose shell sculpture is alike inside and out (e. g., ammonites). Cores of snails, clams, and ammonites are the commonest of all fossils. Among plant fossils, cores are no rarity in the horsetails. The giant equisetids of the Carboniferous period are principally preserved as cores of the pith zone, i. e., the original medullary hollow of the central shoot filled up with sediment (Fig. 4). Such pith cores of other plants are also known, for example, from the cordaitids, a group of Paleozoic gymnosperms. A remarkable type of cores are the natural casts of skull cavities *(cavum cranii)*, usually called "fossilized brains"; these are of the greatest value to the paleoneurologist (see p. 68). The term "fossilized brain" is, of course, misleading as it is not the substance of the brain which has been preserved as a fossil, but merely a filling of the skull cavity. However, this often presents a more or less valid representation of the brain (Fig. 5) and so enables corresponding deductions to be made.

Trace Fossils

Often only the impression or negative of a fossil organism is available, either because the positive (or template) was destroyed during extraction of the fossil, or because there never was anything but an impression, as in the case of tracks. With trace fossils, unlike body fossils, nothing of the original creature remains, merely the traces of its activity. Thus, trace fossils include not only tracks, but also remains of meals, coprolites (fossil excrement), dwellings, traces connected with the processes of reproduction; these give the paleontologist an insight into the mode of life of these early organisms and what they ate.

Miracles of Preservation

The most delicate morphological details can be preserved as fossils in the form of impressions, provided the sediment is sufficiently fine-grained. In this way are preserved traces of organisms which never possessed any hard parts at all, consisting simply of a gelatinous mass,

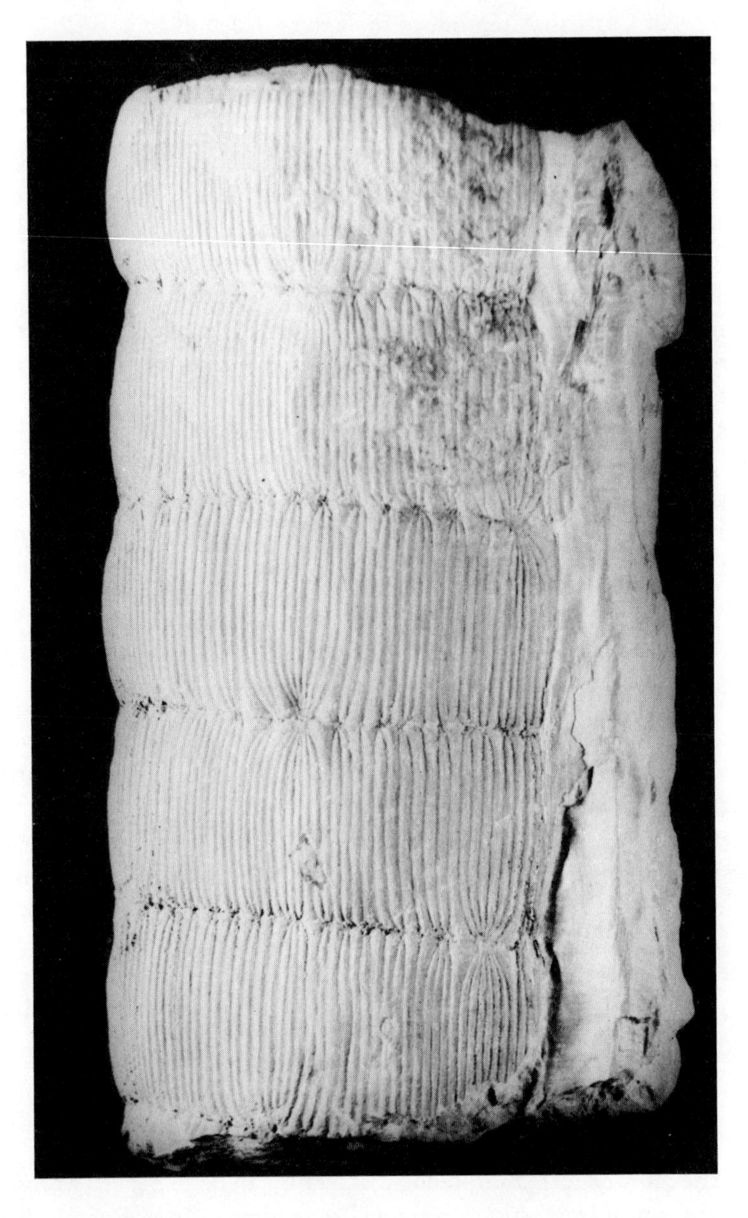

Fig. 4. Core (filling of medullary space) of the stem of a horse-tail *Calamites (Eucalamites) cruciatus* Brongn. from the Upper Carboniferous at Saarbrücken. Much reduced. (After F. Bachmayer, 1957.)

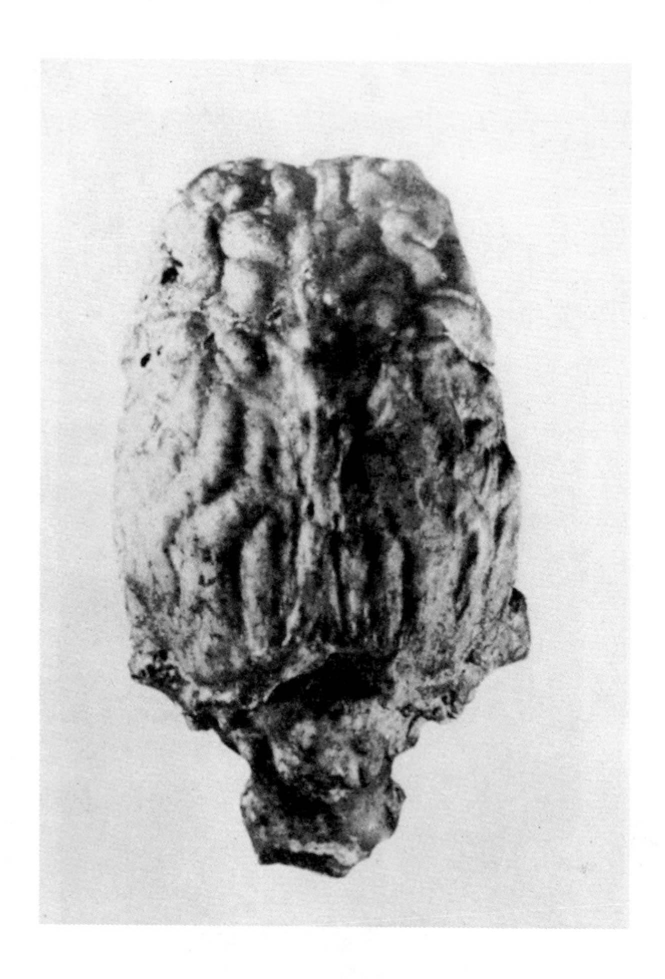

Fig. 5. "Petrified brain" (endocranial filling) of a Pleistocene fallow deer from the Peloponnese. The object looks like a real brain because of the faithful reproduction of the furrows and bumps on the inside of the skull. Reduced.

like the jellyfish. There are some particularly fine impressions of fossil jellyfish from the Solnhofen platey limestones of the South German Upper Jurassic (Fig. 6 c). These impressions actually show the difference in the contraction of the musculature of the umbrella, according to whether it was the upper (exumbrella) or underside (subumbrella) of the jellyfish (Fig. 6 a). Consequently, these jellyfish remains, which all belong to one species, *Rhizostomites admirandus*, have been classified under various species and generic names: *Myogramma speciosum,*

a.

b.

c.

Fig. 6. a and b Recent jellyfish on the beach at Engellöy (Lofoten Islands). a. sunk into the sand in natural position; b. impression of underside. Reduced. (After A. Kieslinger, 1947.) c. Fossil jellyfish *Rhizostomites admirandus* Haeckel from the Upper Jurassic platey limestone of Pfalzpaint bei Solnhofen. Reduced. (After L. v. Ammon, 1886.)

Myogramma speciosissimum, Ephyropsites jurassicus, etc. This mistake was revealed by the observation of A. Kieslinger, based upon comparative studies of Recent jellyfish.

A well-known example of the fossil preservation of fine detail are the remains of the primeval bird *Archaeopteryx lithographica,* found in the same platey limestones. Not only the skeleton of this bird, but even its feathers — at least, their impression — are so well preserved (Fig. 65) that it is possible to see the exact details of the feathering of this species. We find similar examples among the insect remains in these platey limestones, where the most delicate veining of the wings can be preserved (Fig. 7), as can also the wing membranes of the flying reptiles (Fig. 42).

Apart from the impressions we have just been describing, we have so far mentioned only fossil hard tissues. The examples which follow show that, under appropriate conditions, soft tissues as well, right down to delicate and extremely fragile tissues, may be preserved by fossilization. Not only have they survived as impressions but even their physical shape has suffered no significant structural alterations, to an extent which surprises even the expert. This is the reason why we have entitled this section "Miracles of Preservation."

Particularly remarkable are the fossils from the Eocene lignite of the Geiseltal, near Halle an der Saale (German Democratic Republic), which have been described in several publications by E. Voigt. Thanks

Fig. 7. Fossil dragonfly *Urogomphus giganteus* (Germar) from the Upper Jurassic platey limestone of Solnhofen, Bavaria. One-half reduction. Orig. Senckenberg-Museum, Frankfurt/M.

to the excellence of their preservation and the liquid-film method which was used by Voigt, the remains of the soft tissues of various vertebrates could be subjected to examination under the microscope. Fig. 8 a shows epithelial cells of the frog, complete with cell nuclei, while Fig. 8 b shows a melanophore (black pigmented skin cell) of a frog under high magnification. Even microscopically small organisms, such as bacteria, can be recovered as fossils in the Geiseltal lignites, which are at least 50 million years old (Fig. 9).

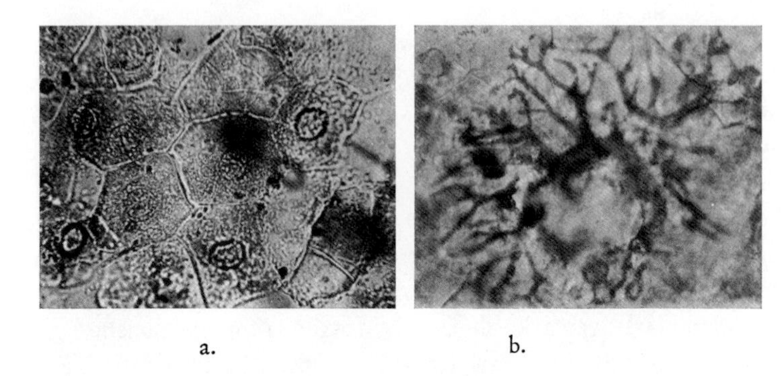

a. b.

Fig. 8. Liquid-film preparation from the Eocene lignite of the Geiseltal, near Halle/Saale. a. Lamellar epithelial cells with nuclei from a frog epidermis × 415; b. melanophore (black pigmented cell) of a frog × 420. (After E. Voigt, 1935.)

The corpses of Pleistocene mammoths from the permafrost regions of Siberia and Alaska are much more widely known. Indeed, the earliest report dates from 1692. To date, about forty of these corpses have been found in Siberia, but few of them were in a relatively complete state of preservation. They are all from the modern permafrost region, and this is how they have been preserved until the present. These mammoth corpses have been the subject of controversy right up to recent times, although the true circumstances were explained many decades ago by O. Herz and E. W. Pfizenmayer. As the fallacies have enjoyed a very wide currency and are constantly being revived by the popular press, let us take a brief look at the reasons for the occurrence and preservation of such corpses.

Most of the frozen bodies were discovered in steep banks of rivers and streams, such as are created or cut back by the annual flooding. The food remnants which have been identified from some of the bodies — mainly true grasses and reeds, very rarely mosses or parts of

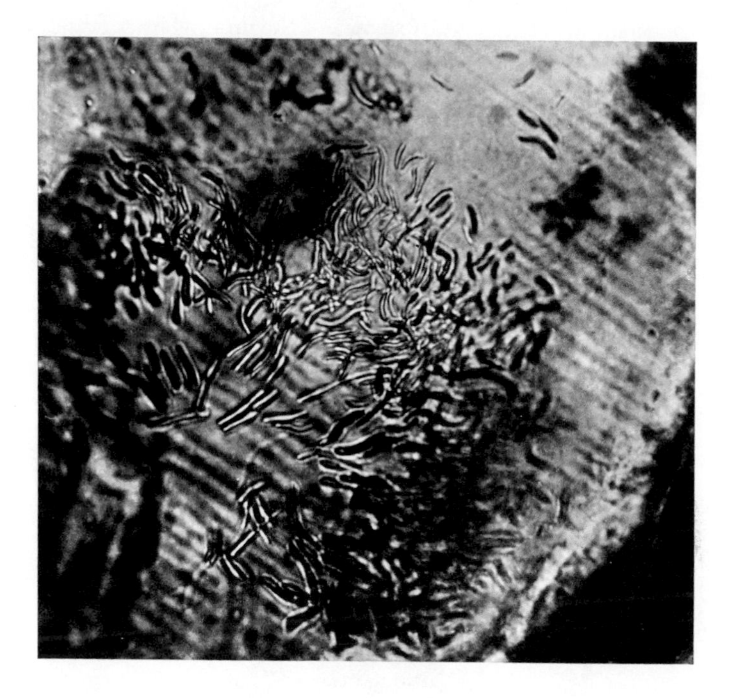

Fig. 9. Fossil bacteria in the trachea of a beetle, *Eopyrophorus*. Middle Eocene from the Geiseltal, near Halle/Saale, ×1000. (After E. Voigt, 1938.)

conifers — tell us that the mammoth *Mammuthus primigenius* lived in a tundra climate. The state of preservation and the position of the various bodies make it quite clear that they were not victims of increasing cold. Most of them were suffocated, either by drowning in the morass or by being buried under landslides thawed by solar radiation. Such places were lethal to such large and heavy animals. A histological examination of the flesh, which looked relatively fresh on the surface, showed that great chemical changes had taken place as a result of slow decomposition, such as occurs with putrefaction. This was supported by the fact that the frozen soil around the bodies always gives off a strong smell of decomposing flesh. This odor is evidence that the mammoth corpses had already begun to decompose before they were frozen, so that no sudden catastrophic change of temperature need be invoked to explain their demise. The bodies of the mammoths were saved from the process of destruction which would otherwise have set in by being embedded in mud. Falling temperature then prevented any further bacterial decomposition and, thanks to the permafrost, the corpses have remained in a state of refrigeration right up to the present (Fig. 10). Outside the permafrost regions, the only remains of mammoths that have been found are the bones and teeth, and especially tusks.

Fig. 10. Right hind leg of a mammoth *Mammuthus primigenius* (Blum.) from Sangayurakh, Siberia, with soft tissues still adhering to bone. (After F. W. Pfizenmayer, 1926.)

Similar circumstances brought about the preservation of the bodies of the Pleistocene woolly rhinoceros *Coelodonta antiquitatis*, described from the tar pools of Starun, near Stanislav in the Western Ukraine, along with those of frogs and birds. It is mainly the extremely thick skin of the woolly rhinoceros which has survived, whereas the horny substances — tusks, hoofs, and hair — have largely vanished. Its preservation is attributed to impregnation by the salt water which was present along with the petroleum, because in tar pools without salt impregnation, no soft tissues of any kind have survived. Asphalt bogs of this type are remarkable for their often massive collections of fossil remains. This richness may be explained by the fact that such asphalt bogs acted as "fossil traps," such as are nowadays brought about artificially in oil-bearing regions. The most famous natural fossil traps are those of the tar pools of Rancho La Brea in Los Angeles, from which several hundred thousand skeletal remains of vertebrates of the Late Glacial and Early Holocene have been recovered (Fig. 11). This is a typical example of a fossil aggregation. Even today, such tar pools still act as "fossil traps," their glittering surface continually attracting birds, insects, and other flying creatures which are then held fast in

Fig. 11. Section of the Pleistocene skeleton accumulation in the tar pools of Rancho La Brea, Southern California. Photo by courtesy of Museum of Paleontology, University of California.

the sticky, yielding surface (Fig. 12). Mammals as well fall victims to the deceptive, clinging tar pools, and they in their turn lure predators to their death.

Further examples of the preservation of soft tissues complete with structural details in fossils are those enclosed in fossil resins, the best known of these being *inclusions in amber*. These are mainly insects (Fig. 13). As Voigt has shown, the very finest details are preserved by the amber — hairs, bristles, muscles, and fibers — although it is only his brilliant method of preparing them by applying a liquid film (see

Fig. 12. Tar pool at Rancho La Brea, Southern California, acting as animal trap. Body of bird with a coating of tough bitumen. Photo by courtesy of Museum of Paleontology, University of California.

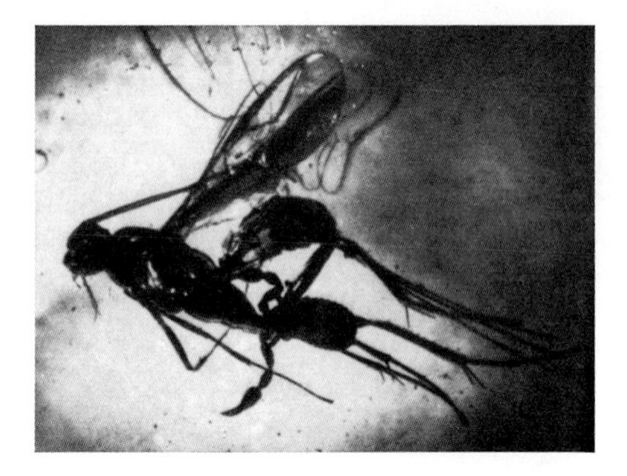

Fig. 13. Pseudoscorpion attached to braconid fly: example of phoresia. Oligocene amber from the Baltic ×12. (After O. Abel, 1935.)

p. 19) which has rendered them accessible to study under the microscope. The rapid engulfing of the objects by the resin prevented the decay of the soft tissues before fossilization. Thus, although it was long believed to be the case, by no means all remains of soft tissues of inclusions in amber are destroyed, although the details are naturally better preserved in small creatures than in large ones.

The Solnhofen platey limestones, which have already been mentioned, have also yielded soft tissues in the form of the phosphatized muscle flesh of fishes, reptiles, and cuttlefish.

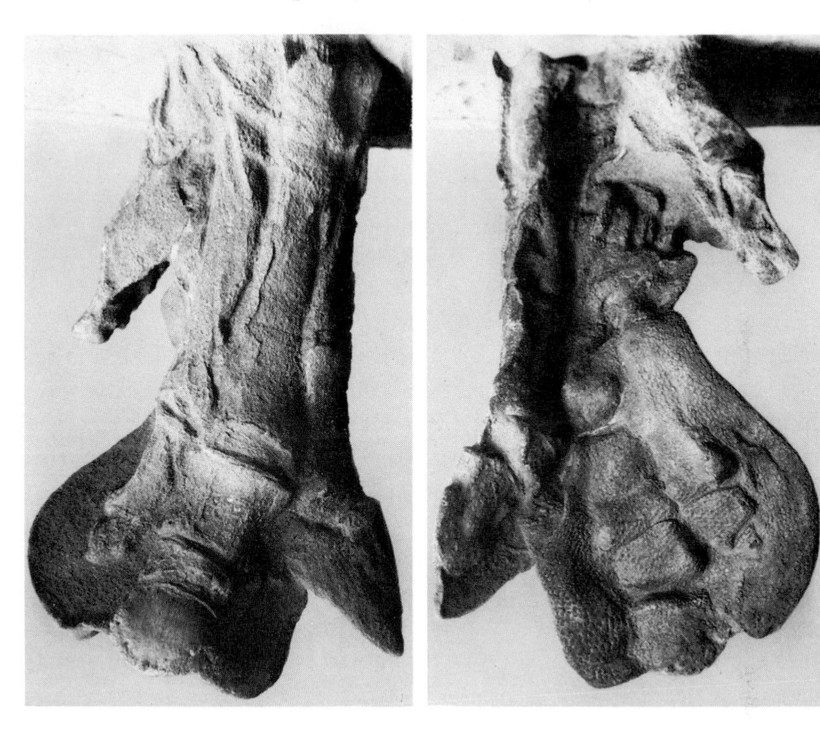

Fig. 14. Right hand of the "mummy" of a fossil duck-billed dinosaur, "*Trachodon*" *annectens* Marsh, from the Cretaceous of Wyoming with fin seam preserved. Front and back. One-half reduction. Orig. Senckenberg-Museum, Frankfurt (Main).

Here, too, we must mention the so-called mummified fossils which have been described as occurring in deposits of various ages. Very often they are assemblages of quite heterogeneous materials which give a deceptive impression of physical preservation. [3] The "*Tracho-*

[3] This is quite apart from those objects which are nothing but concretions, where the remains of plants or animals are encased in more or less thick shells of sediment (e. g., mummiferous limestone from the Jurassic of S.W. Germany=oolite; resin as an embedding material) or where a sort of embalming has taken place, as with opalized woods which have "petrified" in warm springs.

don mummies" from the Upper Cretaceous of Wyoming and Dakota (Fig. 14) are simply pseudomorphs, the body of the animal having

Fig. 15. Fossil mummified frog from the Late Tertiary phosphorites of Quercy, France, Reduced. Photo by courtesy of Musée National d'Histoire naturelle, Paris.

been mummified *before* fossilization; the dried-out organic substance was replaced by inorganic material during fossilization after the mummy was embedded in the sediment, so that the bodily shape of the mummies has been preserved. The "mummies" from the Early Tertiary phosphorites of the French Massif Central (Quercy, etc.) are, according to E. Handschin, true fossils in which the soft tissues — digestive tract, genital organs, etc. — of insects can be studied from polished sections. Apart from insects and their larvae (pupae), frogs are also found as mummies (see Fig. 15). How these came into being is still the subject of discussion, as the hypothesis of the hydrothermal genesis of the phosphorites of Quercy has been disproved on geological evidence, although Handschin considers it must have been necessary for the formation of the "mummies."

Genuine mummified remains are known only from the geologically most recent deposits. We must regard as such the remains of fur and hair of the extinct giant sloth found in North America (*Nothrotherium shastense* from the Gypsum Cave, Nevada) and South America (*Mylodon domesticum* from Ultima Esperanza, Patagonia). [4]

All these finds have been made in areas with an arid climate, this being one of the conditions essential for the creation of natural mum-

[4] An earlier hypothesis, that some of these giant sloths *Mylodon (=Grypotherium) domesticum* were the domestic animals of contemporary prehistoric populations, has not been confirmed.

mies. These sloths lived in the Pleistocene and died out at the end of it, or possibly in the Early Holocene, about 8,000 years before the climate change, according to J. J. Hester. It is remarkable that the coats of the giant sloths *Nothrotherium shastense* which, unlike the present-day tree sloths, lived on the ground, have been found to contain algae of a type similar to those identified in Recent sloths. The remains of muscles and skin recovered from the New Zealand moas cannot, however, be called fossils, as they originate from historic times. On the other hand, unlike the hairs of the giant sloth, the "monkey hair" known to lignite miners has nothing whatever to do with hair. It is fossilized rubber from milk sap vessels, not only from the rubber tree *Ficus elasticus*, but from other trees of the mulberry family (Moraceae) and spurges (Euphorbiaceae), too. The origin of "monkey hair" may be ascribed to "natural vulcanization," obviously a consequence of the sulfur content of the veins.

Some very remarkable occurrences of the preservation of soft tissues of fossil organisms have been discovered in the Posidonia Shales of the Lias of Boll and Holzmaden (Württemberg, Federal Republic of Germany). The fossils from these fine-grained shales are today on display in large museums all over the world. Their state of preservation is extraordinary and includes not only skeletal remains but also, in the bitumen-rich layers (Lias epsilon II, horizons 2–5), soft tissues (skin) which have become museum show-pieces, thanks to the care lavished upon their preparation by B. Hauff and his coworkers. The preservation of the skin of the extinct "fish-reptiles" (ichthyosaurs, see Fig 16) first enabled their exact habitus to be reconstructed.

Fig. 16. Ichthyosaur *Stenoptherygius quadriscissus* Quenst. from the Lias of Holzmaden, with skin preserved. Length of specimen about 2 m. Orig. Senckenberg-Museum, Frankfurt (Main).

Many fossils preserve fragile projections and bristles; or the faceted compound eyes (Fig. 17) made up of numerous single eyes (ommatids) known from trilobites, Paleozoic arthropods; or the delicate spiral loops of some Mesozoic brachiopods (Fig. 18). All these may well be called "miracles of preservation." In these cases, too, rapid embedding

Fig. 17. Faceted eye of a trilobite *Cyclopyge prisca* (Barrande) from the Ordovician of Bohemia. Length of eye 11 mm. (After F. Bachmayer, 1957.)

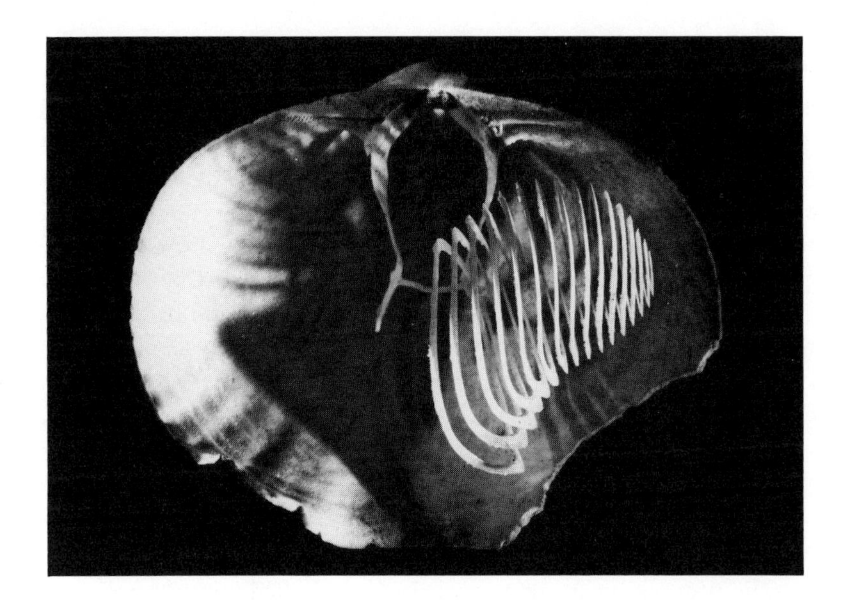

Fig. 18. Brachiopod *Spiriferina pinguina* Ziet. from the Jurassic of France, with the spirally coiled loop preserved on the left side. About one and one half times natural size. (After F. Bachmayer, 1957.)

in a fine-grained sediment is a prerequisite for good preservation. Equally outstanding is the condition of fossils in cherts and siliceous shales where silification has brought about the preservation of unicellular organisms from the Precambrian.

The retention of color in fossil organisms is best known from the shells of molluscs — snails, clams, cephalopods, and brachiopods (e. g., *Lingula*) — as well as from insects in amber. They enable the original color pattern to be reconstructed. Nevertheless, particularly in pre-Tertiary remains, a secondary color change must be taken into account.

Biostratinomy and the Sites of Fossil Discoveries

Some of the examples already discussed have shown how essential it is to have some knowledge of what went on *before* fossilization occurred, in order to determine not only the process of fossilization but also the habitat and way of life of what was once a living creature, be it mammoth, giant sloth, or any other organism.

To explain these phenomena is the task of biostratinomy (Efremov's "taphonomy"), a discipline founded by J. Weigelt, which is concerned with all that happens from the time an animal starts to die until it is finally embedded. A great help here is "contemporary paleontology", [5] which we shall mention again later. Biostratinomy is extremely important for an assessment of the phenomena of fossilization, and this can be decisive when it comes to ecological analysis (see p. 51). There are many questions which arise in this connection — for example, those which concern the alignment of shells, etc., by flowing water such as the direction of transportation and hence the type of flow, the genesis of the so-called Muschelpflaster (mollusc pavement), and places where arthropods shed their skins (exuvia = remains of sloughed skins), which are of value to others as well as the paleontologist. Fossil Muschelpflaster, for example, may help to identify the stratigraphic top and bottom of layers and is thus of practical use to the geologist.

The sites of fossils are also very closely linked with the sphere of interest of biostratinomy. Fundamentally, autochthonous sites — i. e., those in primary deposits — are to be distinguished from allochthonous sites — i. e., those in secondary deposits. Here the questions refer to the period which followed fossilization as well as to the one which preceded it. The question of site is of the greatest importance to the geologist, who uses index fossils to determine the age of the deposits in which they occur.

Autochthonous fossils include those whose habitat and burial place were one and the same, as is usually the case with sessile organisms. [6] Allochthonous occurrence implies a transposition of the fossils, the time of this transposition being very important. If it occurred before fossilization and hence within the same geological time span, it is called a synchronous allochthonous occurrence; if the transport process took place after fossilization, it is called a heterochronous allochthonous occurrence. Index fossils can therefore only be used to determine the age of deposits in the absence of heterochronous allochthony. Such redepositions are not brought about, as the layman usually

---

[5] The term "contemporary paleontology" is contradictory, but has now become accepted.

[6] Tectonically caused relocation of fossils as part of a whole complex of rocks (nappe) is disregarded in the definition of autochthonous and allochthonous fossils.

assumes, by mountain-building processes, but predominantly by flowing water in rivers, lakes, or seas, or by the breaking of waves on the shore; on sea coasts this can result in the breaking up of fossil-bearing layers, with the consequent exposure of fossils and their re-embedding, as shown in Fig. 2. Secondary deposition can occur, for example, when land mammals are found in marine deposits. The paleobotanist distinguishes limnal (autochthonous) from paralic (allochthonous) coal fields because the former have a stump horizon (upright, rooted trunks of tree-like growths, cf. Fig. 48) with a foundation of roots,

Fig. 19. "Petrified forest" from the Triassic of Arizona (Jasper Forest National Monument). In foreground remains of trunks of petrified conifer *Araucaryoxylon arizonicum*. Photo by courtesy of National Park Service.

while the latter are identified by the interpolation of marine layers between the coal seams and dolomite nodules. The "petrified forests" in deposits of various ages found in Arizona, Patagonia, Central Europe, S. W. Africa, and Egypt usually consist of tree trunks washed down into a pile by floods and found in the form of calcified or silicified wood (infiltrated by a calcareous solution or amorphous silicic acid, with deposition in the actual cell or cell walls). Exceptions are the "petrified forest" in the Triassic of Arizona (Fig. 19), where the wood is embedded in bentonites, and the Early Tertiary petrified forest at Gallatin in Yellowstone Park, where fossil trees, upright on their roots, are found in several horizons on top of each other, resembling the stump horizons of coal fields.

As may be seen from Fig. 2, fossils may not only be secondarily exposed and re-embedded, but also deformed by pressure and high temperature, or even totally destroyed (Fig. 20). For the most part, the deformation of fossils caused by sedimentary pressure or tectonic forces is easily recognized. It is, however, more difficult to evaluate

a. b.

Fig. 20. Damaged fossils. a. Ammonite *Uptonia jamsoni* Sow. from the Lias of Hechingen, Württemberg, with damaged living chamber; about one-third natural size; b. coral *Thecosmilia fenestrata* Frech from the Upper Triassic (Rhaetian) of the Zlambachgraben (Upper Austria) with chalice structure destroyed by pressure and heat along a slickenside on a fault plane (left); about one-third natural size. Orig. Palaeont. Inst., Vienna.

fossils that have been deformed without breakage, as, for instance, those in the marine molasse of St. Gallen (Switzerland). It took experimental investigations to prove that the numerous "species" of mollusc were based upon the remains of shells which had been deformed without breaking. Similar distortions are known to have occurred in fossil vertebrates during fossilization.

It is, of course, taken for granted that in all fossil finds of this kind only natural changes are referred to, that is, changes which have not been artificially produced by the hand of man. Among artificial changes are fossil forgeries, which are mentioned here mainly for the sake of completeness. According to the perpetrator and the purpose he had in mind, such fossils look more or less genuine. Usually these forgeries are made by laymen, witness the Beringer Lügensteine. Sometimes they are made by collectors who wish to complete their collection — ignorant of the morphological-anatomical structure they make

Fig. 21. Dendrite from the Solnhofen platey limestone at Kelheim. Reduced. Orig. Palaeont. Inst., Vienna.

several fossils into one. The most successful forgery ever made was the Piltdown man *("Eanthropus dawsoni")* which deceived scientific experts for decades until a fluorescence test some years ago unveiled it as consisting of part bones of recent apes and part fossil bones of geologically recent human beings. There was a time, too, when amber inclusions were manufactured — frogs, lizards, and other animals were artificially introduced into the amber. Usually, the mixed refraction betrays that it is a forgery, or pseudo-inclusion.

Objects are frequently found in nature which look just like fossil remains and are regarded as such by the layman, who often brings them to the expert for identification. These are mostly concretions which may resemble bones in size and shape (e. g., "loess babies" in Pleistocene loess; septaria [concretionary nodules] in Tertiary marine sediments). Then there are dendrites (Fig. 21), which look rather like ferns but are in fact derived from iron and manganese which penetrated the rocks in solution and precipitated out. The layman also may find "Tutenmergel," or silica rings, which have a relatively regular structure and sculpture but are nevertheless of inorganic origin.

III. Fossils and Folklore

Many remnants of primitive forms of life look so different from living organisms on account of their state of preservation that the nonexpert may be excused for not recognizing their true nature. Resemblances to familiar objects have led to false interpretations. These interpretations have in turn led to queer ideas, widespread among simple people, particularly about very remarkable fossils or those which occur in large masses. Such beliefs have been handed down from medieval times — in some cases from antiquity — and they are very persistent, particularly among country people who live closer to nature.

In this chapter I shall mention just a few of the more extraordinary examples which shed some light on superstitions based on fossil finds; in some cases the link between fossils and folklore emerges clearly.

"Petrified Hoofmarks"

In the Triassic limestones of the Alps (e. g., the dolomites of the South Tyrol and the Northern limestone Alps in Austria) there are numerous heart-shaped marks to be seen in the rock surface, which do look a little like the hoofmarks of cattle (Fig. 22). They are often present in large numbers and sometimes reach a diameter of several tens of centimeters. These are the remains of molluscs, called Dach-

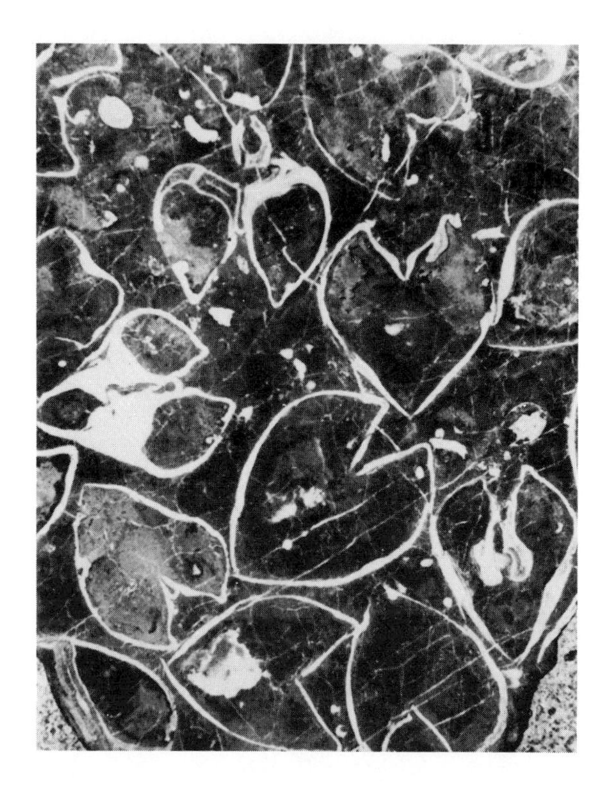

Fig. 22. "Petrified cow tracks" from the Rhaetian Dachsteinkalk of the Lueg pass, Salzburg. Cross-sections of "Dachstein pelecypod" *Conchodus infraliassicus* Stopp. Diameter of cross-sections 12—18 cm. (After H. Zapfe, 1957.)

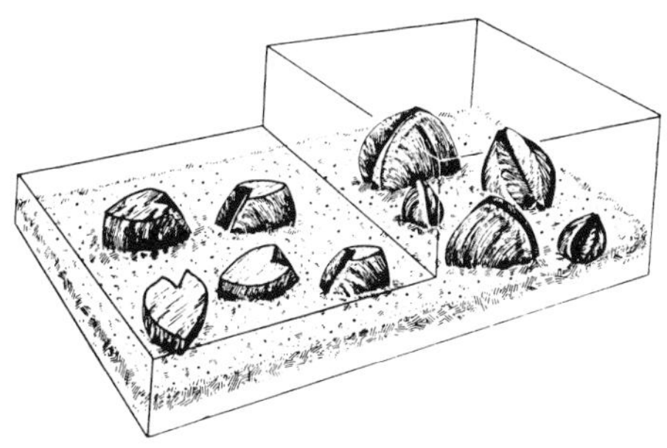

Fig. 23. Block diagram showing "Dachstein pelecypods" in their life position in the limestone mud. On the left the subsequent weathering of the rock surface is indicated to show how the cross-sections came into being. (After H. Zapfe, 1957.)

stein mussels because they are mainly found in the Dachstein areas, or megalodonts (principally *Conchocus infraliasicus*) which are highly characteristic of the Late Triassic (Rhaetian stage). They are frequently preserved as cores, their cross-sections being very different in size and shape (Fig. 23). The Alpine cowherds call them "petrified cow treads," but there are still places where one finds the original explanation of these fossils as "tracks of game animals which ran across the rocks" (O. Abel). Such ideas go back in time to well before the introduction of Christianity.

Petrified "Beans" and "Coins"

Much more common, and hence more easily reconciled with a number of ancient beliefs, are the large Foraminifera (unicellular organisms) belonging to the nummulite group *(Nummulites, Assilina)*, which are often found in petrified form (Fig. 24 a). Very characteristic

a. b. c.

Fig. 24. Fossils and folklore. a. Nummulitic limestone with specimens partly loosened by weathering from the Middle Eocene of the Paris Basin. b. "Petrified money" *Nummulites perforatus* d'Orb. from the Middle Eocene at Bakony. c. "Petrified beans" *Nummulites* sp. from the Upper Eocene of Guttaring, Carinthia. All figures reduced. Orig. Palaeont. Inst., Vienna.

of Eocene deposits, they are shaped like beans or coins, being rounded and sometimes flat (Fig. 24 b, c). The pyramids of Gizeh are built of nummulitic limestone blocks brought from huge quarries in the Mocattam Mountains on the right bank of the Nile. As the stone weathers, large numbers of the rounded nummulites are shed, and thus the ground around the pyramids is strewn with these foraminiferan shells. Their bean-like appearance of course gave rise to the legend that they were fossil beans, left behind by the slaves who built the pyramids. The subject of legends at various other locations, nummulites are generally linked with a story of hasty flight.

Thunderbolts and Sun Wheels

Even more superstitions surround the finding of belemnite remains. These are portions of the inner skeleton of fossil cuttlefish, known as rostra to the scientist, which usually are well-rounded, compact, bullet- or rod-shaped objects, pointed at one end. Frequently found in Jurassic and Cretaceous deposits, these remains are exposed in the course of weathering because they are more resistant than the rock. In the South German Lias they are found in thick, multilayered deposits, as a result of their having been washed down by floodwaters (they have been called "belemnite battlefields"). To quote F. A. Quenstedt: "There is no petrifact which has intrigued the people of South Germany more than the belemnites." They have been called lynx stones (lyncurium, because on rubbing they give off a smell of ammonia, reminiscent of cat's urine), thunderbolts, thunder and lightning stones, devil's fingers, accursed stones, and so on; they have been used as charms to ward off evil, and, in powdered form, as remedies against being struck by lightning or being bewitched, and as cures for a variety of illnesses (Fig. 25 a).

Ammonites are equally fertile in generating superstitions; they are cuttlefish shells, coiled into tight spirals, which often occur as fossils in Mesozoic deposits. According to their shape and condition (for instance, pyritized ammonites have a golden gleam), they are known as golden snails, serpent stones or ophites, dragon stones, etc. (Fig. 25 c).

The fossil "sea lilies" are animals and not plants; crinoids (e. g., *Encrinus liliiformis* from the Muschelkalk), and the remains of their stems are found in many rock-forming layers; these are called crinoidal limestones (e. g., the alpine Hirlatz limestone, Fig. 25 b), and they are known as sun wheels or St. Boniface' pennies. The stem of the "sea lily" is made up of disc-like joints, called trochites, which usually disintegrate before fossilization, revealing the ray-like marks on the inner surface. The Germanii identified them with solar radiation and regarded them as objects of religious veneration. Later, when the heathen beliefs were driven out by Christian doctrine, these sun wheels were "converted" into St. Boniface' pennies.

Fig. 25. Fossils and folklore. a. "Belemnite battlefield" consisting of rostra of *Hastites clavatus* (Quenst.) from the Dogger of Württemberg. b. Crinoidal limestone consisting of stem sections of "sea lilies" from the Upper Carboniferous of Louisville, United States. c. "Serpent stone" from the Lias of Württemberg. The orifice of the ammonite *Coroniceras rotiforme* (Sow.) has been worked into a head to make it look like a coiled snake. (After F. Bachmayer, 1958.) All figures reduced.

Dragons' Bones and Giants: Stories of Dragons and Fossils

No less interesting are the stories which have grown up around the the fossil remains of vertebrates. In olden times it was not unusual to find the tusks and marrow bones of the extinct mammoth *Mammuthus primigenius* in Late Pleistocene deposits (loess and river debris), and these were thought to be the relicts of human giants.[1] Abel argued

[1] The real ancestors of man were, as we know from fossils, rather smaller than present-day Europeans. The Pleistocene fossils, supposed to be the remains of giants, are either those of primates—e. g., *Gigantopithecus* from China—or they are those of men who, though they had larger teeth and massive jawbones (*"Meganthropus" palaeojavanicus, "Pithecanthropus" modjokertensis, "Zinjanthropus" boisei*), had a smaller overall body size than most modern Europeans.

that the tales of the monster Polyphemus in Homer's Odyssey can probably be attributed to discoveries of fossil skulls of the dwarf elephant *Palaeoloxodon falconeri*, etc. found in caves along the coast of Sicily. Abel thinks that the notions of a one-eyed monster Polyphemus, or the Cyclops, can be explained by the fact that the nasal orifices in the front of the skull had widened into a single opening, and this was supposed by persons ignorant of anatomical proportions to be a single eye socket (Fig. 26). Although in this example the con-

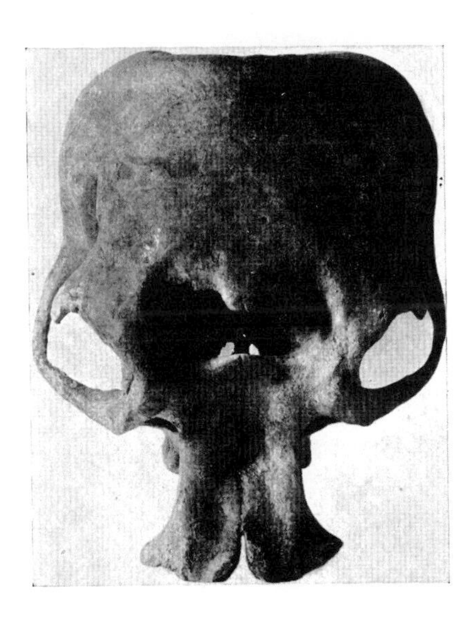

Fig. 26. Pleistocene dwarf elephant skull *(Elephas falconeri)* from the cave of Spinagallo, Sicily, frontal view. The tusks are missing. Note median nasal aperture that was interpreted as an eye socket, thus giving rise to myths about Cyclops (one-eyed giants) in the Mediterranean area. Much reduced. (After B. Accordi, 1962.)

nection between fossil discoveries and legend may seem somewhat tenuous, in the case of the Lindwurm monument in the main square of Klagenfurth, Carinthia, Austria (Fig. 27 a), the story is fully attested. As was recognized by F. Unger, Ulrich Vogelsang, the sculptor who created this monument, begun in 1590, used as his model for the dragon's head the calvarium — skull minus lower jaw — of an Ice-Age woolly rhinoceros *Coelodonta antiquitatis* found in the Lindwurm quarry to the north of Klagenfurt about 600 years ago (Fig. 27 b).

Abel has also attempted to produce evidence to the effect that the belief in dragons was not based solely upon the skulls of Ige-Age mammals — for example, the numerous skulls of cave bears from the Dragon Cave near Mixnitz in Styria, Austria — but also upon other fossils of vertebrates, namely the remains of extinct reptiles, such as the plesiosaurs of the South German Lias beds, of which complete skeletons have been found.

a.

b.

Fig. 27. Fossils and dragons. a. The Lindwurm monument in Klagenfurt, Carinthia. b. Calvarium of Late Pleistocene woolly rhinoceros *Coelodonta antiquitatis* (Blum.), possibly from Zollfeld near Klagenfurt, supposed to have been used as a model for the head of the Lindwurm. a. Picture postcard. b. (After O. Abel, 1939.)

Such superstitions have arisen quite independently among various peoples, as may be surmised from the beliefs of populations long settled on continents other than Europe. The bones of the rhinoceros-like titanotheres (= brontotheres), exposed after heavy rainfall in the Badlands of Nebraska and Dakota, were known to the Sioux Indians as the remains of "thunder horses," while the Navajo Indians of Arizona held that the weathered silicified tree trunks of the "petrified forests" (see p. 29), often broken up into pieces, were the bones of huge giants (Fig. 19). "Dragons' bones and teeth" belong to the stock-in-trade of Chinese apothecaries; these are for the most part the fossil remains of mammals from Pleistocene cave deposits or from the widely distributed fossil-rich red clays of the Chinese Pliocene. These bones and teeth play an important part in Chinese medicine. The ancient Chinese also regarded the mammoth corpses from the Ice-Age deposits as the bodies of underground-burrowing animals which were doomed to die upon exposure to the light of day. Futhermore, the Pampas Indians of Patagonia thought the skeletons of huge mammals found in the Pleistocene glacial loess of the pampas showed that there had once been huge beasts living in underground lairs.

These few examples make it clear that the ideas about fossils in folklore were many and various and that discoveries of fossils were indeed at the root of many legends and did contribute to a belief in dragons.

IV. Working Methods in Paleontology

Technical and Scientific Methods

A distinction is usually made between technical and scientific working methods. Both methods begin with the discovery of the fossils and end with scientific evaluation, either in the form of publication or with the the display of fossil objects in museums. In practice, there is no strict division between the two since they run side by side and scientific evaluation begins (from the point of view of biostratinomy) as soon as the fossil is uncovered.

While we shall not attempt to give a systematic account of all the methods used in paleontology, we have chosen a few examples to provide some insight into the great variety of working methods and analyses.

Collection, Digging Out, and Excavation of Fossils

Depending on where the fossils are found — almost without exception in sedimentary rocks — one need only pick them up (in fields, vineyards, etc.), dig them out (in sand or gravel pits and quarries), or finally mount an organized excavation (in caves, fissure-fillings, etc.). Excavation may sometimes require prior removal of the overburden by a bulldozer, or perhaps trial digging in the form of small pits. One excavates the layers of sediment, one at a time; prepares a grid map and sketches the position of the fossils (skeletons or single bones) for each horizon excavated, and then assigns a number to each horizon (Figs. 28 and 29). The plan will be used later on in the laboratory to reconstruct the original positions of the various remains. This is equally important for the prehistorian who wishes to know whether there are any traces of the cultural practices of paleolithic man (for instance, skull burial) and for the paleontologist who can determine from the plan whether the deposit was brought down by flooding.

Fig. 28. Excavation site in the Grenzbitumen horizon of the Anisian stage of the Triassic, point 902, at Monte San Giorgio in Ticino Canton, Switzerland. Alternating beds of dolomite and bitumen. (Photo E. Kuhn-Schnyder, 1961.)

The many difficulties which may confront a paleontological excavation include a high water table making it necessary to pump out the site, problems of access to the diggings, or the need to traverse cave systems.

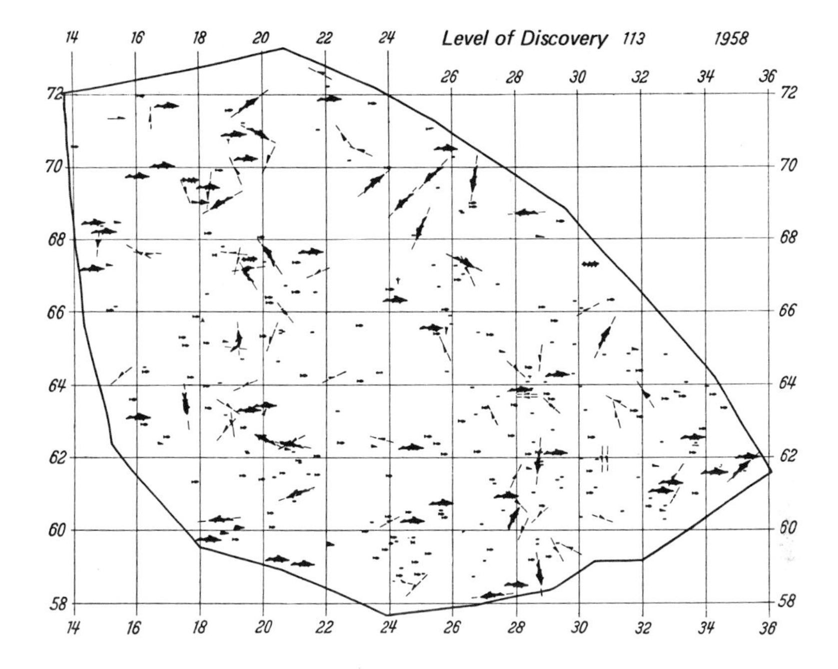

Fig. 29. Plan of discoveries in layer 113 of the site shown in Fig. 28. Abundant ichthyosaurs—*Mixosaurus,* fish—*Saurichthys, Birgeria,* and coelacanths, coproliths and plant remains. The plan shows the position of the various finds. (By courtesy of E. Kuhn-Schnyder, Zürich.)

Preparation of Microsamples and Microfossils

The recovery or excavation of macrofossils is matched in micropaleontology by the taking of sedimentary rock samples which contain microfossils[1] (as hand samples or as cores) and preparing them for examination. This procedure, which is essential in micropaleontology (including palynology, the analysis of pollen and spores, see Fig. 31) is carried out in the laboratory; its object is to separate the fossils from the sediment. Sometimes it is merely necessary to wash them well on a sieve, meshes from 0.5 to 0.15 mm being used to separate different

[1] The designation "microfossils", together with the extremely minute nannofossils, includes all fossils which can be examined only under a microscope, irrespective of their systematic classification, e. g., foraminifers, radiolarians, ostracods, conodonts, otoliths, silicoflagellates, diatoms, coccoliths, pollen and spores. However, since this is also true of the larger Foraminifera—*Nummulites, Assilina, Lepidocyclina, Fusulina, Schwagerina*—these, too, are considered microfossils (cf. Fig. 30).

Fig. 30. Microfossils—foraminifera and ostracods—from the Upper Rupelian clay (Oligocene) of the Mainz Basin. ×21. (After K. Krejci-Graf, 1955.)

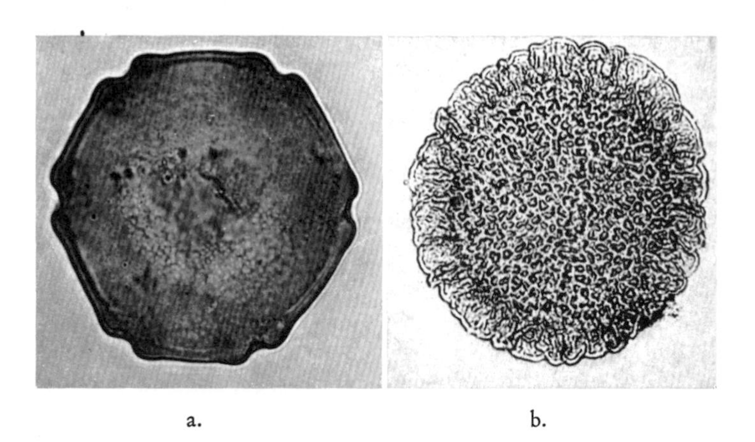

Fig. 31. Pollen grains from the Late Tertiary. a *Pterocarya* from Köflach, Styria, ×1000. b. *Tsuga* (hemlock pine) from Hausruck, Upper Austria, ×500. (After W. Klaus from E. W. Petrascheck, 1956.)

fractions from which the microfossils are then sorted out under the stereomicroscope; sometimes it is possible to separate the soil residues from the microfossils by means of their specific gravity by the use of a heavy liquid like carbon tetrachloride. However, it is often necessary

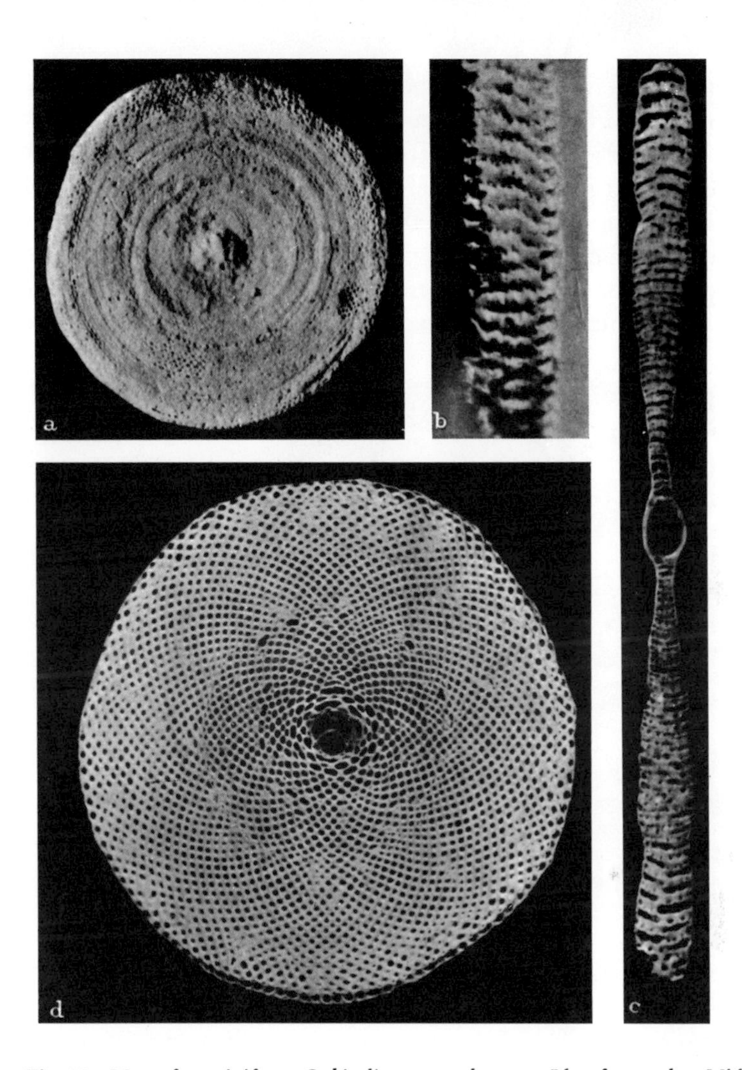

Fig. 32. Macroforaminifera *Orbitolites complanatus* Lk. from the Middle Eocene of the Paris Basin; a. view of shell, ×12; b. section of the marginal side, ×24; c. axial section, ×24; d. equatorial section, ×18. (After R. Lehmann, 1961.)

to apply special recovery methods, either because the sediment cannot be washed off readily, or because it has already hardened into rock. In such cases, the following preparations will help to remove the sediment: hydrogen peroxide, Glauber's salt, hydrochloric acid — for noncarbonized microfossils in limestones; strong caustic soda — for radiolarians and unicellular organisms with a silicate skeleton; hydrofluoric acid — for spores and pollen in clays. Sometimes, as with the tiny radiolarians, it is merely necessary to etch the surface, since they need not be freed completely from the sediment. With silicate cherts, a similar effect may be obtained by moistening the sample with oil, as this makes the rock appear transparent to a depth of several millimeters, allowing the enclosed microfossils to be examined optically. Many rocks have to be broken up mechanically first if the fossils, e. g., conodonts, cannot be extracted by means of chemicals, e. g.,

a. b.

Fig. 33. Electron microscope photographs of Recent and fossil coccoliths. a. *Coccolithus huxleyi* (Lohmann) from a bottom core of the South Atlantic, ×15,000 (After M. Black and B. Barnes, 1961). b. *Coccolithus sarsiae* (Black) from the Late Tertiary of the West coast of England, ×3600. (After M. Black, 1962.)

monochloracetic acid. It is important that the microfossils in the sediment shall not be attacked, and possibly destroyed, by the solvent. The basic materials of microfossils are calcium carbonate, silica, sporonin, and pollenin.

After they have been sorted under the stereomicroscope, the microfossils may be systematically isolated for further treatment in what are known as Franke cells [2]; stored in Plummer cells as a collective fauna preparation, or placed on a watch glass and embedded in Canada balsam, Palatal (a polyester resin), Caedax, or any other

[2] Franke cells are discs of plastic or cardboard, a few millimeters thick, usually having a circular, black-painted depression; they are used to store microscopic objects and may be closed by a cover-glass, thus enabling the objects to be stored loose inside the cell without the necessity for fixing them.

suitable medium, so as to form permanent microscopic preparations. With large Foraminifera, thin sections are cut in several directions after they have been embedded in a polyester resin, e. g., Polestar (see Fig. 32). This procedure illustrates the difficulty mentioned at the

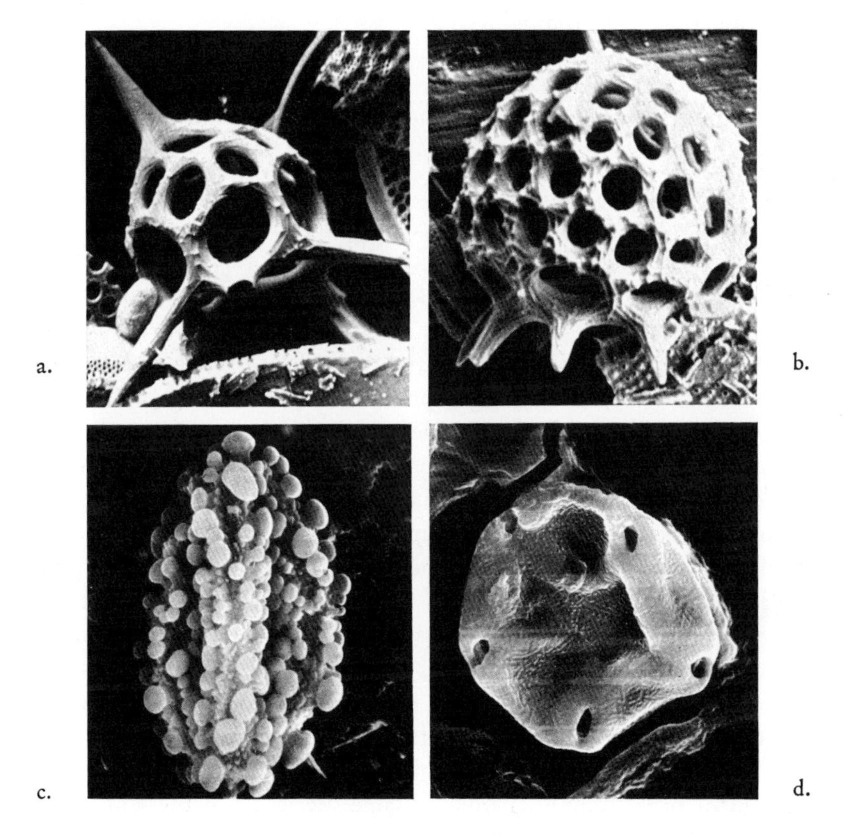

Fig. 34. Stereoscan photographs of microfossils. a. *Cannopilus ernestinae* and b. *Cannopilus sphaericus;* silicoflagellates from the Upper Miocene of Santa Barbara, California, ×750. (After A. Bachmann and A. Keck, 1969). Pollen grains from c. *Ilex aquifolium* and d. *Alnus* sp. from the Riss-Würm interglacial of Mondsee, Upper Austria. ×935. (Photos W. Klaus, Vienna.)

beginning of this chapter of separating working techniques from scientific methods. The electron microscope has recently been used to reveal the microstructure of nannofossils, which is too minute to be distinguished under the light microscope; the smallest nannofossils are less than 40 microns = 0.04 mm in size, for example, coccoliths. After this procedure the taxonomist can go to work on them (Fig. 33). The scanning electron microscope (Stereoscan) has proved particularly valuable here, because its depth of focus enables both ultrastructure and ultrasculpture to be distinguished (Fig. 34). Moreover, its field of application is not limited to microfossils.

Conservation, Preparation, and Mounting of Macrofossils

Macrofossils are conserved and prepared, depending on their state of preservation and site of recovery, either immediately upon discovery or after they have been taken to the laboratory. Very fragile fossils which would not stand transporting, require soaking and fixing with such preservatives as shellac, cellulose, and Geiseltal lacquer at the spot where they are found. Vertebrates, after prior fixing, require encasement in a plaster cast reinforced with fabric bandages, because the entire complex of finds, complete with the surrounding sediment, has to be transported in one or more blocks, depending on size, to the workshop for preparation. Here the cleaning, preparation, and restoration of the skeleton is carried out and then, if desired, mounted for display in a museum. Mounting not only requires that the complete skeleton of a fossil vertebrate be available but also that the person responsible for the mounting has sufficient zoological training to make it look life-like. There are, however, problems in displaying extinct vertebrate animals — their posture and manner of locomotion can only be guessed at. Incomplete skeletons require the reconstruction of missing or fragmentary bones, either by reference to the remains of other individuals of the same species or, if there are none available, by reference to other closely related species.

Sometimes the value of a fossil object lies almost exclusively in how it is prepared. Some of the best-known fossils have already been mentioned (p. 35) — reptiles, fishes, "sea lilies," and so on — from the Posidonia Shales of Boll, Holzmaden, and Ohmden in Württemberg, which are showpieces (Figs. 16 and 17). These have only become objects worthy of display because of the painstaking technique of preparation used; at the same time they have been made accessible to scientific study. Before their preparation, even the splendidly preserved examples of the various reptiles were just shallow elevations in the appropriate overburden which betray to the expert alone the presence of fossil remains. The preparation, which is usually done under the stereomicroscope, requires not only skill in the use of hammer, chisel, and steel burin but also a knowledge of the anatomy where it is not elucidated by X-ray photographs before the prepa-

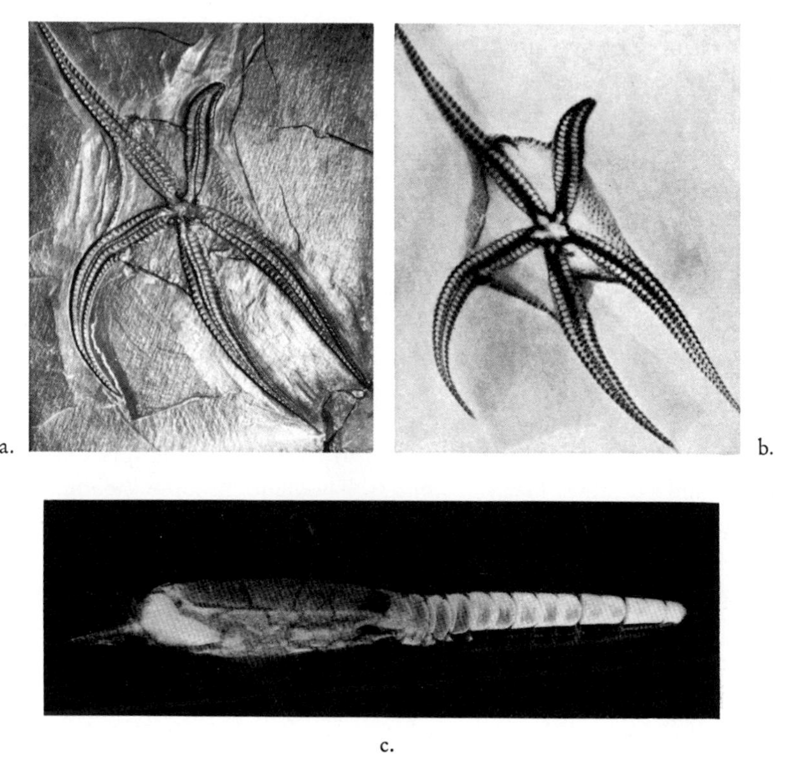

Fig. 35. Starfish *Encrinaster (Aspidosoma) roemeri* (F. Schöndorf) from the Upper Devonian of Bundenbach, Rhineland Schiefergebirge. a. Exterior photograph of the prepared specimen; b. X-ray photograph. Reduced. (After W. M. Lehmann, 1938); c. Cephalopod *(Lobobactrites)* with soft tissues from the Middle Devonian of Wissenbach, Harz. X-ray photograph, slightly enlarged. (After W. Stürmer, 1969.)

ration is done. The Swiss paleontologists B. Peyer and E. Kuhn-Schnyder are justly famous for the preparation of fossils from the Grenz-bitumen horizon of the Triassic from Monte San Giorgio in the province of Ticino (Switzerland); also noteworthy are the preparations of the starfish and arthropods from the Hunsrück Shales of the Rhenish Lower Devonian of Bundenbach and the remains of fishes and plants from the Mansfeld Shales of the Thuringian Permian. It is not usually sufficient to split these finely layered shales in order to expose the fossils. Generally the successful preparation of such remains requires the use of fine wire brushes and sand blasting machines, since the fossils are often silicified and hence more resistant than the actual shales. The use of X-rays is important, not only for the subsequent preparation of the fossils (Fig. 35), but for the discovery of fossils, too. Thus, among other things, the very delicate extremities of trilobites and also the original holothurian (sea cucumber) from the Devonian shales of the Hunsrück were found by means of X-ray photographs and were thus prepared in a life-like manner. Very fine silicon sulfide crystals are formed from which it is possible to identify the original soft tissues on the X-ray picture (Fig. 35). In some cases, ultraviolet light may also be used to locate fossils.

Another technique is the preparation of serial sections and their purely morphological evaluation, rather like the wax-disc method used in zoology and medicine. It was first used by the Swedish paleontologist Erik A:son Stensiö in 1927 to examine early Paleozoic Agnatha, the jawless creatures distantly related to the modern lampreys and hagfishes (= cyclostomes). The agnatha have a completely ossified skull capsule (Ostracodermata) and, thanks to the serial-section method, we know more in some respects about their anatomical structure than about the Recent Cyclostomata, since it has been possible to reconstruct all the details of their central nervous system and blood-vessel system (Fig. 36). The sections were cut at very close intervals, which, of course, resulted in the total destruction of the fossils; but this was compensated for because they were later available as prints or as wax-disc models of the various cross-sections and could be put together to form a whole. In certain favorable cases, it was even possible to make casts.

Recently serial sections, which are also used on fossil invertebrates (e. g., brachiopods), have been applied to the preparation of cellulose peels. Here the sections are etched with hydrochloric or hydrofluoric acid, and the subsequently applied film generally displays more detail than the polished section. It also provides a record of the original before it is destroyed by sectioning.

Another field where much intensive work has been done in recent years is *paleohistology*, a branch of paleoanatomy concerned with the study of the microscopic structures of the hard tissues. Paleohistology has provided valuable information on the composition and formation of the components of bones and teeth. Here, too, thin or polished sections are required.

With fossil woods (silicified or calcified wood, or carbonized lignite woods, known as xylite), thin or polished sections cut across or along the specimen (radial or tangential) are necessary to assess the anatomical structure of the vascular bundles (trachea, tracheides, medullary rays, resin pathways, etc.).

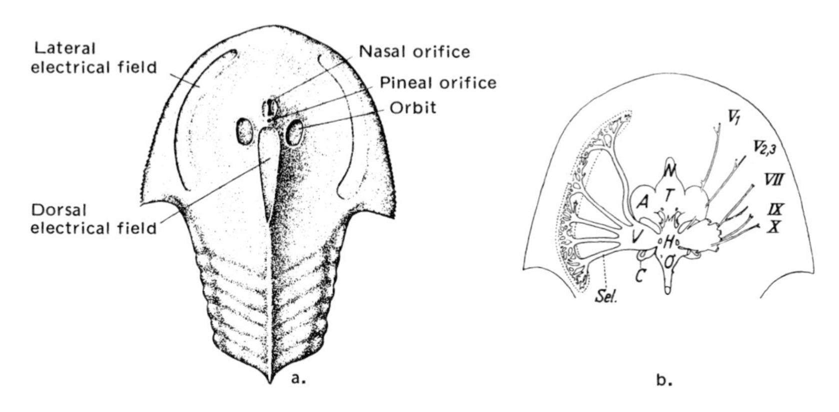

Fig. 36. Cephalaspid *Kiaeraspis auchenaspidoides* (Stensiö) from the Upper Devonian of Spitzbergen. a. Carapace from above; b. cast of skull cavities from below. Artificial casts of (A) eye capsule, (C) ear canal, (H) rear brain, (M) mid-brain, (N) nose capsule, (O) sensory field, (T) forebrain, (V) vestibule of hearing organ, (V_1, $V_{2,3}$, VII, IX, and X)—nerve canals. (After A. S. Romer and E. Kuhn-Schnyder, 1953.)

Another paleobotanical method, called cuticle analysis, is carried out on the remains of leaves. The lacquer film method is essential for their preparation; this was first used by Voigt for the recovery and conservation of the soft tissues of fossil organisms (see p. 18). The subject of cuticle analysis is the outer skin of the leaf which is preserved in fossil form in the carbonized leaf residues, making it possible to form some idea of such taxonomically important structures as the stomata, the cells of the epidermis and other features essential to systematic studies (Fig. 37).

Not only mechanical but also chemically acting methods are employed to prepare microfossils, particularly when the rock is harder than the specimen and mechanical preparation might damage the fine details. Thus, etching with acetic acid is used for the phosphatized remains of the vertebrates flattened into the layers of the Solnhofen platey limestones; this, with their simultaneous embedding in plastic resin, has proved an excellent method.

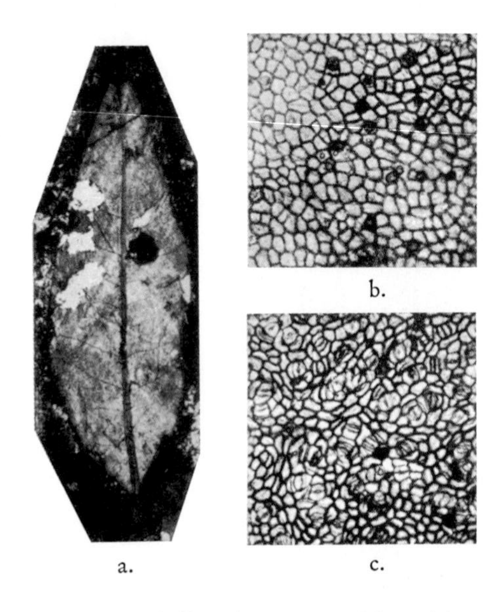

Fig. 37. Leaf of *Laurophyllum bournense* with cuticle preserved from the Eocene lignite of the Zeitz-Weissenfels district, GDR. a. Leaf, reduced two-thirds. b. epidermis of the upper surface of the leaf, ×60; c. epidermis of underside of leaf, ×60. (After K. Mägdefrau, 1968.)

Scientific Studies According to Taxonomic, Chronological, Chorological, and Phylogenetic Criteria

Once the fossil remains have been prepared, they are ready for scientific study. In principle any distinction between paleozoological and paleobotanical methods is impossible. Both types of study are directed toward the same end, the only difference being in the different nature of the discipline's objects, or their degree of preservation.

Similarly, the special features of micropaleontology do not derive from the manner in which the specimens are studied scientifically so much as from the methods applied to the preparation of specimens.

The scientific study seeks to achieve not only as exact a knowledge as possible of ancient forms of life, but also to arrange them systematically according to when they appeared and their distribution on the Earth, to explain their phylogenetic links and, finally, to get some idea of the relationship between these creatures and their environment, both animate and inanimate.

The systematic or taxonomic classification finds its expression in the species nomenclature. After a taxonomic assessment, it is possible to attempt to position a fossil according to period or stratigraphy, i. e., chronologically. A systematic and chronological interpretation is usually preceded by what is called an ecological analysis, although this frequently begins with the exposure of the remains, since the place and circumstances of the final position of the fossil remains relative to each other and to the sediment, and so on, may be of material importance.

Taxonomy, as a classifying science, attempts to fit all living things into a system. Such a system may be either artificial or natural. There are many difficulties, which will be discussed later, that allow only an artificial system to be constructed. However, the aim of science is to achieve a natural system, dependent upon the interrelationships of organisms and the origin of species. Hence, the natural system must remain faithful to the evolutionary events. The system in use today was devised by Linnaeus who, in the eighteenth century, undertook to arrange all the then known Recent plants and animals into one grand system; this involved giving each one two names, one denoting species and the other genus, thus creating the foundation for an internationally accepted nomenclature. Linnaeus still relegated fossils to the mineral kingdom (lapidum regnum). Because of the relatively few characteristics adduced as criteria, the Linnean system was in many respects artificial, as could only be expected at the time when he lived. Even today, science is still a long way from achieving a natural system.

The systematic evaluation of fossil remains often encounters great difficulties. Some of the reasons are that the evaluation deals with utterly extinct creatures, which have no equivalent in the modern flora and fauna, with remains that are in an imperfect state of preservation, and with remains that are in isolation. This is true virtually only of fossils with articulated hard parts (e. g., vertebrate skeletons and echinoderms and arborescent plants) which were separated from their natural association before fossilization, or which when they were still alive — at least, for a time — no longer had a link with the remainder of the organism, as is the case with pollen, spores, fruits, seeds and leaves.

The following examples may help to give some idea of the difficulties which can arise; let us first take a few from the field of paleobotany. The palynologist, in particular, has to overcome numerous obstacles which arise from the state of preservation. The botanist who performs pollen analyses — investigations of peat bogs, determinations of the origin of honey, and studies of glaciers by means of pollen analysis — can determine by direct comparative studies to which varieties of plant isolated pollen grains belong. The paleobotanist cannot do this, because only in extremely rare cases have pollen-bearing organs (anthers) been preserved in fossil form. He is therefore entirely dependent on comparisons with Recent pollen grains if he wishes to make a determination. This method works for Pleistocene pollens and spores, as Pleistocene plants are identical with Recent ones. But it fails completely with isolated pollens and spores of the Tertiary or earlier, sporae dispersae, as they are called, which are not found in association with the sporangia, and which originate from what are today tropical and subtropical species, or perhaps from totally extinct ones. The degree of similarity between pollens and spores has nothing whatever to do with the degree of affinity of the species to which they belong. Therefore it is impossible to make a systematic arrangement of isolated pollen and spores unless they can be unequivocally identified with Recent pollen and spores. As a result, purely artificial systems are resorted to, by means of which some order can be introduced among the sporomorphs, as pollen and spores are called. Here the units, unlike species, genera and higher systematic categories of the natural system (the taxa), are named by shapes of species and shapes of genera (parataxa) and are treated differently from the units of the natural system from the point of view of nomenclature, too. The systematic arrangements of nearly all fossil sporomorphs within particular species and genera — and hence their botanical affiliations — remain a closed book to us; even so, there are no limits to their use in stratigraphic correlation. What is important here is not only botanical origin but also ready recognition of the various index forms and the ability to distinguish them from other forms which may look alike.

The species is the basic unit of the system. Linnaeus, originator of the system of binomial nomenclature, gave each species a specific name and a generic name, e. g., *Canis lupus,* the wolf. All systematic units above species — genus, family, order, class, phylum — are mere abstractions. Zoologists and botanists define the species as the breeding community. Every individual in a population which is interfertile, i. e., can breed together under natural conditions and produce fertile progeny, belongs to the same species. Such a definition of species is no use at all to the paleontologist, for its essential criterion is physiological, and one that he is unable to judge. Therefore his concept of species is fundamentally morphological. The basic prerequisite for defining a species in the paleontological sense is the recording of all morphologically important characteristics, taking into consideration as large a series of individuals as possible so as to get an idea of the range of variation. This definition leaves a great deal to the feeling for systematics of the scientist making the evaluation. This fact makes it easy to understand why scientists have such widely differing views, which may lead to variation in the limits of individual species. The outcome of such systematic investigations is ultimately expressed in species names.

In contrast to taxonomy, which is the science of classification, nomenclature is built up according to internationally accepted rules and is a technique which simply serves to provide the species or any other systematic unit, as defined by the scientists, with an internationally intelligible label; in cases of doubt, the oldest name is the valid one, since any more recent names for the same species are simply synonyms and hence invalid. Thus, the generic names once used to describe various states of preservation of arborescent lycopsids, such as *Bergeria, Knorria,* and *Aspidaria,* are all synonyms of *Lepidodendron,* as it has since come to be appreciated that the various "genera" are all in effect lepidodendrons in different states of preservation. *Stigmaria* and *Lepidostrobus,* too, are no more than parts — roots and fruiting organs — of lepidodendrons, and are no longer used as genera in the true sense of the term. True, it is often not easy to determine which species they do belong to in cases of this kind. It is frequently some chance discovery which reveals the association.

The paleobotanist is in a similar situation when working with fossil woody plants found in isolation in the form of remains of wood, leaves, fruits, seeds, and pollen. This can and does lead to a situation where one species is described under three distinct names without any mutual identification being possible. It is therefore usual to speak of wood flora, leaf flora and pollen or spore flora.

Cuticle analysis, as mentioned above, is very important in the identification of fossil leaves, since a purely morphological analysis of the leaf — shape of the whole, outline, leaf edge, rib pattern, and stem joint — is often inadequate for a determination of species, at least, where leaf remains of the Tertiary and beyond are concerned,

a. b.

Fig. 38. Graptolites from the Silurian of Czechoslovakia. a. *Monograptus priodon* (Bronn) from Reporyje, about four-fifths natural size. b. *Monograptus turriculatus* (Barr.) from Litochlav, about three-fifths natural size. Orig. Naturhistorisches Museum, Vienna. (Photos F. Bachmeyer, Vienna.)

quite apart from the fact that complete leaves are necessary for this, whereas this need not be so for cuticle analysis. Experience has taught us that the leaves of one plant may be of widely differing shapes whereas the leaves of plants which are not particularly closely related may be very much alike (convergence phenomena). The characteristics determined by cuticle analysis are much more varied than the shape and form of the leaf.

Now let us take a few more examples from the field of paleozoology. First a group of animals which is completely known.

For a long time small shining objects, usually only a few centimeters long, have been found in Silurian shales (Fig. 38); some have a toothed edge on one side, reminiscent of a fretsaw blade, some have it on both sides, and they occur either singly or in the form of a rolled spiral or in "bundles." These are graptolites and they play a very important role as index fossils in helping to classify the Ordovician and the Silurian. A closer study of these objects revealed that they are colonies of animals whose individual members are housed in small cups fixed along a central axis. The skeleton consists of chitin, which is familiar to us from insects. For a long time the systematic position of these organisms was controversial, although they are fairly easily distinguished from other groups and well differentiated even among themselves. They were initially supposed to be plants; then they were thought to be related to the Hydrozoa and Bryozoa. But investigations in recent years, using improved methods of preparation, have shown that they are related to the Pterobranchia, a small group of animals, generally unknown save to the zoologist. (Pterobranchia almost always occur in colonies and the exoskeleton, like that of the graptolites, consists of half-circles of chitin contiguous to one another along zigzag seams.) This discovery determined their taxonomic classification, so today the graptolites are regarded as an extinct group within the Stromochordata.

In contrast to the graptolites, even a refined method of investigation has not led to any elucidation of the systematic-phylogenetic position of the conodonts. These are microscopically small tooth-shaped objects (Fig. 39) found in Paleozoic and Triassic deposits. At first they were thought to be real teeth, but histological studies have shown that their growth proceeded in the reverse direction to that of teeth. The direction of growth of the conodonts and the material they are made of show that they cannot be the teeth of vertebrates or even invertebrates — for instance, of the segmented worms known as scolecodonts. The conodonts are presumed to have been excrescences on the skin of groups of Chordata, now wholly extinct, which had no other hard parts.

The taxonomic assessment supplies the basic data not only on the distribution of a species in space (chorology) and time (biostratography) but also on its abundance and phylogeny.

The biostratigraphic analysis of fossils which help the geologist determine the relative ages of fossil-bearing sedimentary rocks, also begins with the recovery of the fossils. To be able to date rocks, we need *index fossils,* that is, fossils characteristic of a particular span of time, such as a biozone, stage or longer unit of time (see p. 74).

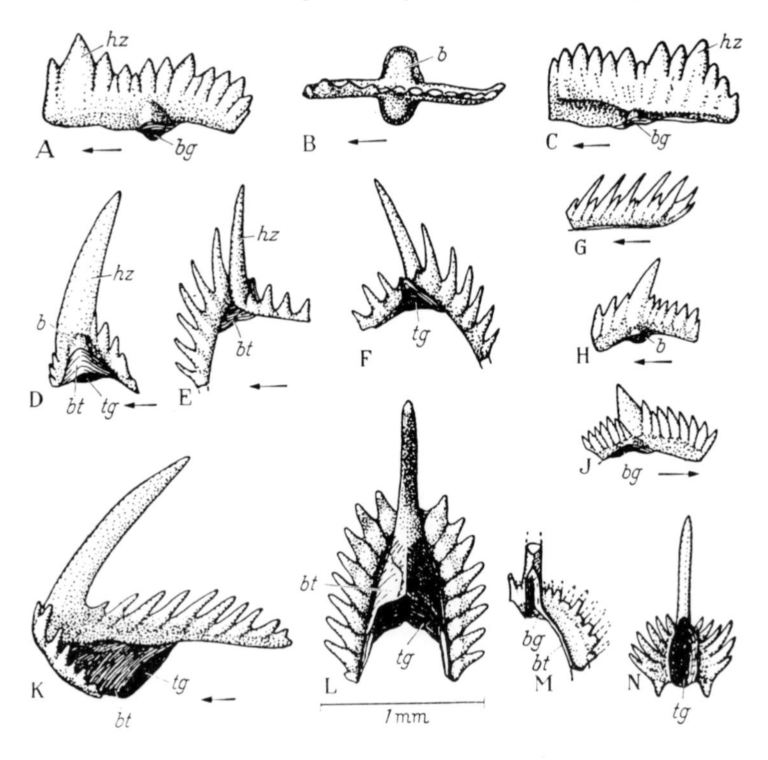

Fig. 39. Conodonts from the Silurian. A—C *Ctenognathus,* D *Prioniodina,* E and F *Lonchodina,* G *Hindeodella,* H and J *Ozarkodina,* K *Ligonodina,* L—N *Trichonodella.* (After W. Gross, 1957.)

It is important to recognize when fossils have been moved after fossilization and hence are found in geologically younger deposits. Proof of a large-scale experiment of this kind by nature are the countless "erratic" blocks which were carried during the Ice Age by the continental ice of Scandinavia down to North Germany, complete with all their fossil content.

Apart from the limitations on the interpretation of age by means of index fossils due to the nature of their sites, there is yet another source of error. This source, which is biologically determined, is the facies, i. e., the aspects of the sediment and its organic content which are determined by local peculiarities, such as whether they are lake or marine deposits. Facies differences may be interpreted as differences in geological age, and vice versa. The biostratigraphic classification of the Late Tertiary beds of the Vienna Basin is based less upon the evolutionary changes in the various groups than upon ecologically determined alterations, the initially exclusively marine deposits giving way in the course of time to brachyhaline and brackish and finally to fresh-water deposits. The purely marine character of the fauna in the deposits of the Vienna Basin in the earlier "Tortonian" period (*Bulimina, Bolivina* and *Rotalia* Zones) already shows some impoverishment in the later "Tortonian," suggesting that the Vienna Basin had been isolated in the meantime. In the Sarmatian this isolation was certainly complete. The rich variety of species of marine fauna gives way to a small range of species, although this was compensated for by a larger number of individuals in each. This process of substitution can be followed step by step and, as it is basically an ecologically determined change, it is a reflection of one more of nature's large-scale experiments; accordingly, it admits of a fine stratigraphic subdivision *(ecostratigraphy).*

It will, however, be clear that facies differences can introduce complications when parallels are drawn between deposits of similar age. Thus, in principle, only flora and fauna of similar facies — known as isopic facies — should be compared to avoid false conclusions in drawing parallels as to age. The differences which can be found in every lake or sea between the flora and fauna of the basin and littoral species are quite sufficient. Solving problems of this kind is really the task of chorology, with an ecological analysis at the same time. The word "paleoecology" has recently come into use, meaning basically what used to be called paleobiology. However, paleobiology has been used rather loosely; on the one hand in Abel's sense to mean simply the study of the manner of life of fossil organisms, and on the other hand to mean the whole of paleozoology and paleobotany (zoology plus botany = biology), which made it in effect merely a synonym of paleontology, as was pointed out by L. Dollo, R. Richter, and O. H. Schindewolf. Thus, it is better to avoid the term altogether and to speak rather of paleoecology and paleoethology (cf. p. 64). The true founder of paleoecology is the Swiss naturalist O. Heer, as he was the first to employ paleoecological principles in connection with his work on the Oeningen fossils (see p. 143).

Reconstruction of Fossil Organisms

Before going into paleoecological analysis, there is a little more to be said about the *reconstructions* which are usual in paleontology. The paleontologist is often asked whether reconstructions of extinct animals have a sound scientific basis. There is no question that the production of reconstructions is the proper concern of paleontology; the important point is to be aware of the limitations of scientifically based reconstructions and to resist the temptation to go beyond them or to lay false claim to knowledge that is not, in fact, available.

Reconstruction methods are for the most part applied to extinct vertebrates. This means not so much the mounting of skeletons but rather the *reconstruction of parts which have not survived in fossil form,* in effect, the soft tissues. These include both the musculature and the covering of the body — fur, scales, feathers, etc. This type of reconstruction must be based on a complete skeleton and appropriate information about the nearest related living form. Fig. 40 illustrates the method of an exact reconstruction of an extinct vertebrate, the reconstruction of the skeleton being followed by that of the muscles, thus forming the basis for the reconstruction of the complete body. The closer the relationship with a living species, the easier this process is. For large animals of the Late Ice Age the cave drawings of paleolithic man provide valuable supporting evidence, for the artist of that age really knew his game animals and generally depicted them in a most lifelike manner (Fig. 41). We thus know the external appearance of these forms. It is precisely the external appearance of extinct vertebrates about which exact information is lacking — with rare exception — because the paleontological record is scant (see p. 13 ff.). The type of covering, the formation of the mane in mammals, the feathering in birds, the nature of the body surface, skin folds and crests in reptiles and amphibians — these are the details most open to criticism in reconstructions of extinct vertebrate animals. The reconstructions of reptiles, prehistoric birds, and "Stegocephalia" (primitive Paleozoic and Triassic amphibians) of the sort which appear mainly in popular scientific works are often extremely highly colored. Viewed with scientific objectivity, the coloring is rather imaginative; this does not

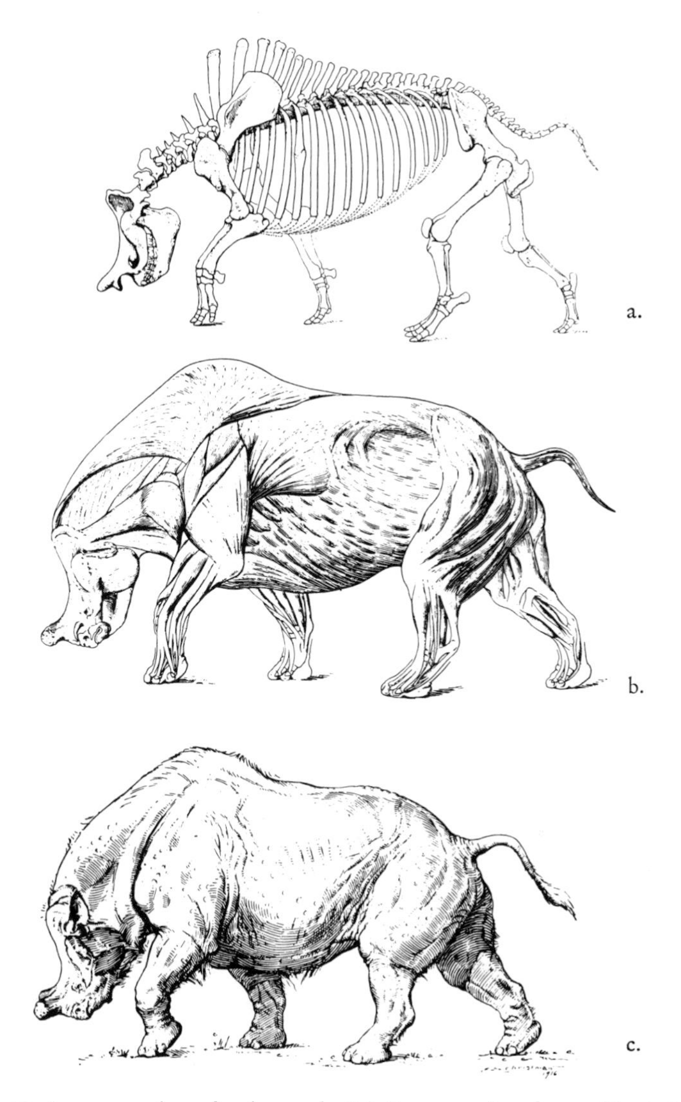

Fig. 40. Reconstruction of a brontotheriid *Brontops* cf. *robustus* Marsh from the Early Tertiary of Dakota, United States. a. Skeleton; b. surface musculature; c. habitus. (After H. Osborn, 1929.)

mean that it is quite unknown for color to be preserved in fossil organisms (cf. p. 27). From another viewpoint, indirect conclusions may be drawn about the characteristics and peculiarities, not preserved in fossil form, of long extinct animals which lack close relatives among

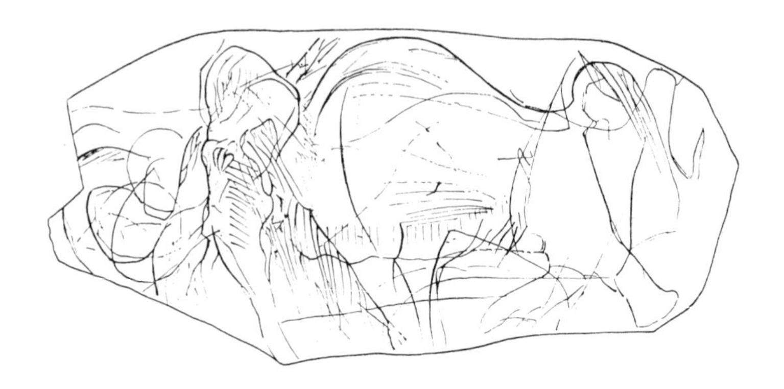

Fig. 41. Mammoth *Mammuthus primigenius.* Drawing scratched on a fragment of tusk by a Late Paleolithic man; from La Madeleine, Dordogne. Length of fragment 245 mm. (After E. Lartet from O. Abel, 1927, redrawn.)

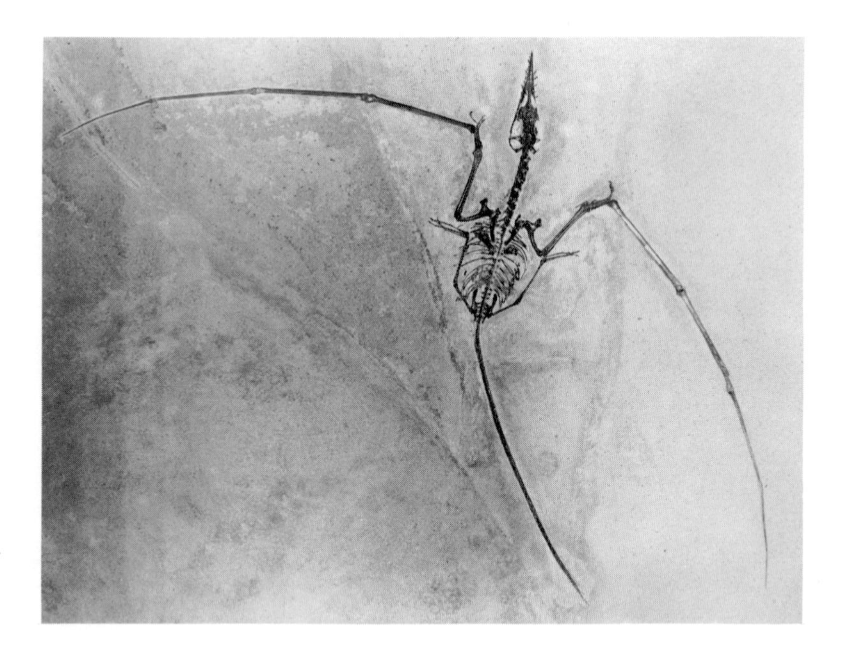

Fig. 42. Pterosaur *Rhamporhynchus gemmingi* H. v. Meyer from the Upper Jurassic platey limestone of Solnhofen. Rhomboid tailplane recognizable in outline. About one-eighth natural size. Orig. Senckenberg-Museum, Frankfurt (Main).

the modern fauna. For instance, we may assume that in the ancestors of mammals among the Reptilia (Therapsida), known mainly from finds of skeletons in Triassic deposits, the presence of ethmo- and naso-turbinalia in the nose cavities indicates not only that they were warm-blooded but also that they had a hairy covering. Because of these and other peculiarities of the skeleton, these creatures are known in the literature as mammal-like reptiles (cf. p. 94).

Much the same applies to the completely extinct flying reptiles (Pterosauria), numerous more or less entire skeletons of which have been recovered from Jurassic and Cretaceous deposits. These reptiles had a membrane for flying which was opened by means of a much elongated fourth finger. In addition to the short-tailed species (Pterodactyloidea), there were also long-tailed forms with a horizontal tail plane (Rhamporhynchoidea); because there are several examples of these in an excellent state of preservation, we know what they looked like (Fig. 42). There are no really comparable forms among the modern fauna, as Recent reptiles include no specimens capable of flapping flight. The flying dragon *Draco volans* merely glides with the aid of a sail of skin extended sideways by a prolongation of its ribs, and the biologically comparable "flying fox" is a mammal which differs, too, in the structure of its flying membrane. In particular, the nature and manner of the flying reptile's locomotion on land and its resting position have been much discussed by paleontologists. Abel thinks that the resting position of many flying reptiles must have been similar to that of the "flying fox."

The paleobotanist, too, has to make reconstructions to get some idea of the habit of growth and external form of a plant. He is confronted with possibly even greater difficulties than the vertebrate specialist, mainly as a result of the state of preservation. This state often leaves in doubt the original interrelationship of the various parts of the plant, particularly with tree-like growths. Suffice it to mention the arborescent lycopsids *(Lepidodendron)* of the Carboniferous, whose roots were christened "*Stigmaria*" and shoots "*Lepidostrobus*." Eventually, however, the dichotomous forking of roots and branches was established from more complete examples. Our increasing knowledge is reflected in the successive reconstructions (Fig. 43) which also involve paleoecological problems.

To take another example, the pteridospermatophytes (seed-bearing "ferns"), which are completely extinct, look exactly like ferns but,

a.

b.

c.

Fig. 43. Changing fashions in reconstructions of European Carboniferous (= Pennsylvanian) forests; a. after Saporta, 1881; b. after Kukuk; c. after a diorama in the Field Museum, Chicago. (a. and c. after R. Kräusel, 1950; b. after W. E. Petrascheck, 1956.)

unlike them, they were true seed-bearing plants in their anatomical structure and manner of reproduction. They are characteristic of Carboniferous flora. Tracing the history of research on pteridospermatophytes, we find that, at first, all that was known were a few fossil fronds of typical fern structure (but without sporangia) which, however, were attached along axes whose anatomy was completely different from that of a fern. Since the anatomy was more reminiscent

Fig. 44. The Silurian Ocean and its inhabitants. Reconstruction of the fauna from the Budňan Beds near Prague. With cephalopods *(Orthoceras* and *Cyrtoceras)*, trilobites *(Aulacopleura* and *Cheirurus)*, corals *(Favosites, Omphyma* and *Xylodes)*, gastropods *(Murchisonia* and *Cyclotropis)*, brachiopods *(Conchidium)* and "sea lilies" *(Scyphocrinites)*. The colored markings on the cephalopod shells are like those of the originals. (After J. Augusta and Z. Burian, 1956.)

of the Cycadales ("palm ferns"), which were seed-bearing plants, they were called Cycadofilices. At the same time, there were well-known objects which were undoubtedly seeds, although there was only one genus, *Cordaites,* which was known with certainty to produce seeds. Later, more complete examples confirmed the initially suspected connection between the plant with the fern-like leaves and the seeds found only in isolation. It was characteristic of seed-bearing "ferns" not to have true flowers, the seeds being mostly borne on special fronds.

The reconstruction of the individual animal or plant, however, is not the end of the story. The final objective of the paleontologist is to reconstruct what he calls *pictures of life,* that is, the reconstructed individual in his contemporary environment (Fig. 44). And this brings us back to paleoecology.

Paleoecology, Paleophysiology, and Paleoneurology

The aim of paleoecological research is to establish the relationships between the fossil creature and its environment, both animate and inanimate, in the form of life pictures, about which we shall have more to say in Chapter VIII. Thus, unlike stratigraphy, or rather chronology, which seeks to discover connections in *time,* paleoecology concerns connections in *space.* This results in the points of contact we have already mentioned. But paleoecology also takes in the analysis of trace fossils. Since Chapter VII is devoted exclusively to trace fossils, we shall simply mention here some fundamentally important analyses to illustrate their potential as well as to indicate the numerous sources of error to watch out for during paleoecological investigations.

Ecology, which is concerned with living plants and animals, begins with associations or biocenoses, so-called life communities. Only very rarely does the paleoecologist have the good fortune to find fossil biocenoses. He must first laboriously reconstruct their associations, the life communities of long ago, from what are known as *oryctocenoses* (buried fossil communities).[3]

As may be seen from Fig. 45, a basic distinction must be drawn between autochthonous and allochthonous fossil occurrences (see p. 28). This means that there is a close link here with biostratinomy or taphonomy, those branches of science which have to do with the events that occur between the death of the animals or plants and their final embedding (in the sediment); for a paleoecological analysis, the events leading up to fossilization are of crucial importance. At the same time, moreover, it is essential to be thoroughly familiar with the ecological requirements, or way of life, of Recent plants and animals.

---

[3] The term more often used, taphocenosis, refers only to *recently* buried communities and these, not having the capacity for endurance of fossils, are not equivalent to oryctocenoses; this also applies to thanatocenoses, or communities of the dead.

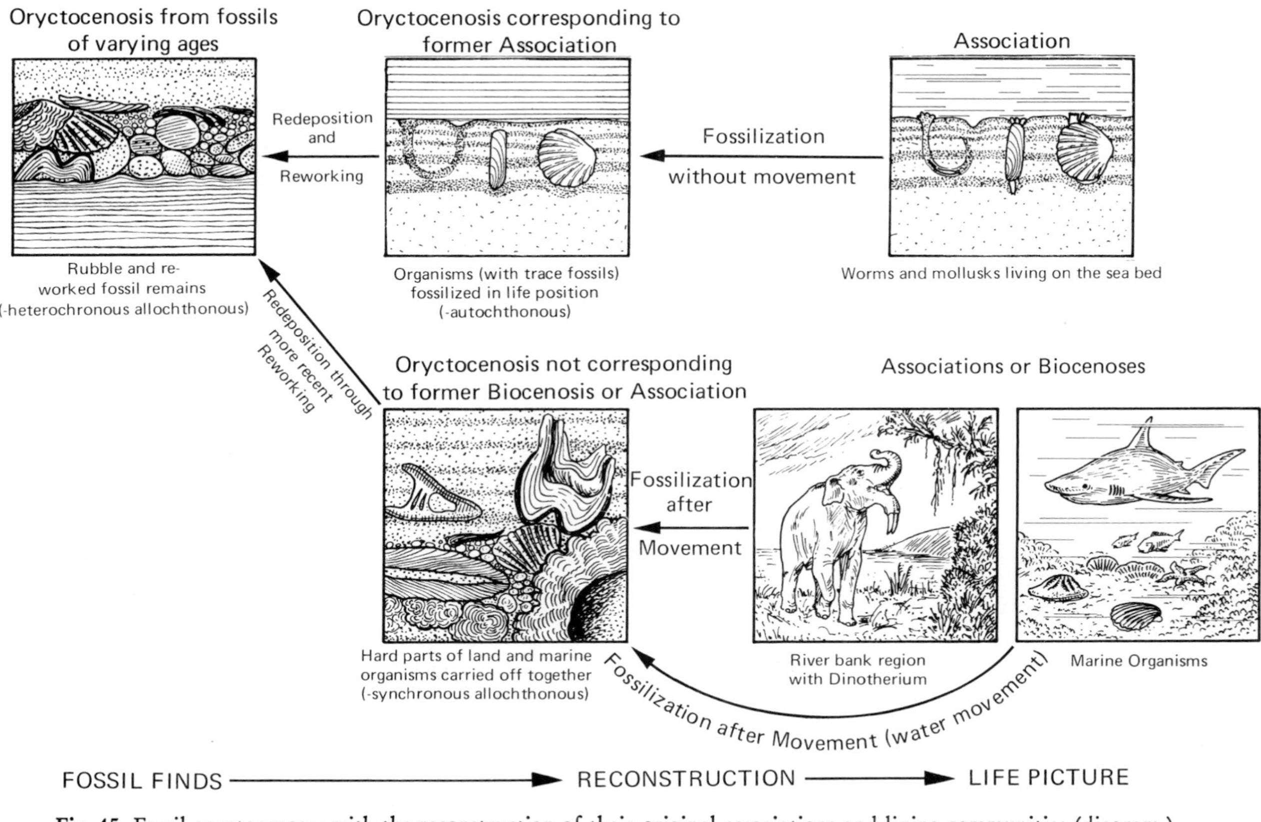

Fig. 45. Fossil oryctocenoses with the reconstruction of their original associations and living communities (diagram.)

Since, however, the questions involved are generally marginal to the preoccupations of zoologists — and botanists (see also Chapter VII), it is necessary to collect relevant observations by making comparative studies on Recent forms of life. This is what R. Richter calls "*actuo-paleontology*" or "contemporary paleontology," though it might be more properly described as "experimental paleontology." True, it does not fit within the framework of paleontology in terms of either material or methods, but it has made decisive contributions to the elucidation of biostratinomic questions (cf. p. 15, on the fossil jelly-fish from the Solnhofen platey limestones).

A fossil animal's way of life can usually be deduced from examining its physical structure and comparing it with Recent organisms. In accordance with the principle that every creature shows signs of adaptation to its environment, these signs are usually the more striking the closer the link with the environment. In addition, both the sediment in which the fossil remains are found and any traces of life which can be discovered must be taken into account. From the physical conformation it is possible to deduce whether the creatures were planktonic, nektonic, or benthonic — floating, free-swimming, or bottom-dwelling in the open sea, respectively — living on the ground, dwelling in trees, burrowing, and so on. But the nature of the find is also important for an evaluation of the way of life. The dentition may provide clues to the kind of food the animal ate (whether it was predator or prey, carnivorous or herbivorous, etc.). There are, however, only indirect clues to certain ecological features, such as temperature, climate, depth of water in which the creatures lived. In recent years *geothermometry* has made it possible to establish many absolute data on the temperature of the water at the time of the formation of the fossils by measuring the isotopes of oxygen (ratio of ^{16}O to ^{18}O) in the calcium carbonate in the skeletons of fossil organisms (e. g., the rostra of belemnites, the shells of Foraminifera, the opercula of ammonites, the skeletal remains of corals, etc.). They are thus called "fossil thermometers" or "carbonate thermometers." Apart from the assumption of normal salinity for the sea water (35‰), there are many possible sources of error to be excluded; these concern the growth of the creature's shell and the depth of the ocean it inhabited.

The "normal" use of fossils as indices of climate relies almost exclusively on comparisons with Recent related forms. For example, palm, cinnamon, and camphor trees and some laurel species — as well as reefbuilding corals which cannot survive where the average temperature of the coldest month falls below 22° C — are assumed to indicate a tropical to subtropical climate. Dwarf willows, dwarf birches, and the snails *Succinea oblonga elongata* and *Arianta arbustorum alpicola* characterize an Arctic or Alpine climate, and so on. The skeletal remains of vertebrates are not very informative about climatic conditions. Here one can only draw conclusions from the type of food (e. g., forest and steppe elephants were browsing and grazing animals, respectively) and, once again, from Recent related forms. The same applies to evaluations of the depth of water in which fossil forms once lived. For example, the current active growth of coral reefs takes place at depths not greater than 46 m; *Lithothamnium* (coralline algae) and barnacles *(Balanus)* live in the surf region; phosphorescent sardines (Scopelidae = Myctophidae) are inhabitants of the deep ocean, etc. All these are conclusions which cannot safely be deduced from the sediment alone. [4] Quite recently trace fossils have been adduced as bathymetric indicators, as shown by A. Seilacher (see p. 121). When using this type of indirect evidence, however, it must be borne in mind that either the ecological requirements or the way of life may have changed in the course of time. Thus sessile (fixed to one spot) "sea lilies" (crinoids) are today found predominantly in fairly deep seas, whereas the fossil crinoids were mainly inhabitants of shallow seas. A similar change has occurred in modern deep-sea crabs, various deep-sea fishes (Macruridae, etc.), and deep-sea molluscs *(Neopilina)* whose fossil forebears are found in shallow-water deposits or even in paralic (seashore) formations. Similar considerations apply to most of the Arctic and Alpine mammals whose ancestors lived at the end of the Tertiary period or in the initial glacial stage in warm-to-temperate climate zones (e. g., the arctic fox, polar bear, wolverine, elk), or were steppe-dwellers or rock-dwellers (e. g., alpine hare, snow mouse, alpine chamois).

Thus the ecological assessment of geologically very old flora and fauna is no easy matter, quite apart from the fact that we have so far implicitly assumed the *principle of actualism*. To the present day it still has not yet been reliably explained under what climatic condi-

---

[4] Certainly, in some cases fairly definite deductions can be made, but these are the exceptions; they include channels, raindrop pits and desiccation cracks due to drying out in marine sediments which show that it was within the range of tidal ebb and flow.

tions coal-forming forests existed. There are indeed some indications that the climate was a tropical one, but they are not sufficient. Another problem is presented by various morphological characteristics of the lepidophytes (arborescent lycopsids) of the Pennsylvanian period which are found only in xerophilous plants, i. e., those which like dry conditions: needle-shaped leaves, number and arrangement of the stomata, which are important in transpiration, and so on. According to their occurrence and association, however, lepidophytes thrived only in markedly humid places (swamp forests). The seeming paradox can, however, be explained by the anatomy of these plants. In contrast to modern woody plants, the proportion of the woody system, which serves among other things to convey water from the roots to the leaves, amounts to a maximum of 15% of the total cross-section, the rest being made up mainly of bark substance. The lepidophytes lacked the distinguishing feature of coniferous and deciduous trees, the secondary growth in thickness of the wood, so that the vital function of support had to be performed by the enormously thick bark. Because of the structure of the water-conveying portion, the lepidophytes suffered from a physiological drought even where water was readily available.[5] We are now dealing with a range of questions which in the terminology used here may fairly be called *paleophysiology.*[6] F. Nopcsa's original definition of this term was, indeed, applied in a rather more restricted sense, for Nopcsa was primarily interested in endocrinology and how it must have regulated the body size of fossil vertebrates and hence their evolution. Nonetheless, the paleontologist, too, can make deductions about the physiological features of extinct animals with a high degree of probability. Let us just mention the problem of homoiothermy (warm-bloodedness) which, on the fossil evidence, was a characteristic not only of mammals and birds, but of their ancestors, the therapsids, as well (see p. 94).

Yet another aspect of paleontology, known today as *paleoneurology* (T. Edinger), should be mentioned at this point. The paleoneurologist is concerned with the central nervous system of fossil creatures and hence principally with their brain. The brain may be examined, even for fossil vertebrates (see p. 13), from either natural or artificial

[5] The xeromorphic character of bog plants may also be connected with nitrogen deficiency.

[6] Paleophotobiology, as J. L. Wilser called it, is also concerned with questions of paleophysiology.

casts of the interior of the skull (cavum cranii) so providing a basis for conclusions. Thus, the brain of the flying reptile was constructed along bird-like lines and was particularly specialized for vision, whereas the earliest known bird, *Archaeopteryx* from the Jurassic, which was a relatively poor flapping flier, had a reptilian type of brain. In interpreting brain structures from fossil endocranial casts, however, some caution is advisable. Thus, Edinger's assumption (based upon well-developed "hearing centers" in a fossil brain) that even in the earliest Tertiary period (Paleocene) bats were twilight insect

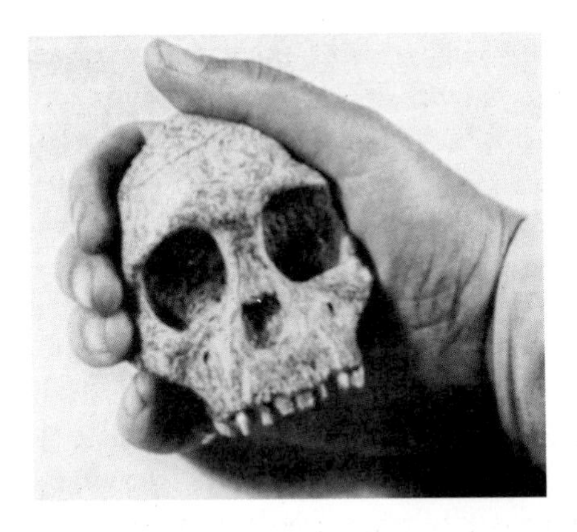

Fig. 46. *Australopithecus africanus* (Dart.). The first fossil remains — skull of an australopithecine child from Taung, South Africa. (After G. H. R. v. Koenigswald, 1968.)

hunters, orienting themselves by means of ultrasound, turns out not to have been justified. And this brings us back once more to paleoecological analysis.

The following example, from the field of vertebrate paleontology, is designed to show the importance of paleoecological analyses not only for paleobotany but also for paleoanthropology. The paleoanthropologist, who is concerned with research into the early forms of man, has available as the basis for his studies only the remains of skeletons and teeth and jaws.

In 1925, when the first such find (the skull of a child — see Fig. 46) was made of an australopithecine, as it was called, the experts were by no means unanimous as to its proper classification; furthermore, its geological age was uncertain because it was found in a fissure filling.

The fundamental question was: Was it an ape or a human being? It became possible to answer this question, which was indeed no easy matter on the basis of a single immature specimen, only after 20 years or so when more complete specimens of australopithecines came to light. Today we know considerably more about these creatures, as there have been a number of additional finds, but still not all the elements of the skeleton are known. In brain size, or rather skull capacity — since this is all that can be measured and this was taken at that time as decisive in assigning specimens to the hominids, i. e., men — the australopithecines are somewhere between Recent apes (chimpanzees and gorillas) and Peking man *(Sinanthropus pekinensis* or *Homo erectus pekinensis).* From a systematic point of view, the dentition — which is very significant — is clearly human (hominid) in form and shows substantial differences from those of all apes. With other findings and the geological age (earliest Pleistocene) this finally led to the classification of the australopithecines among the hominids; the form of the skull and the size of the brain were regarded as less decisive in this case than the upright posture (bipedy) which rests upon the evidence of the shape of the pelvis. As remains of the pelvis show, the australopithecines already walked erect and on two legs in the manner of modern man. This decision not only cleared up the mystery as to why the extraordinarily primitive-looking skull slope of Trinil man *("Pithecanthropus erectus" = Homo erectus erectus)* from Java was combined with an apparently completely modern thigh bone (i. e., corresponding to that of present-day man, *Homo sapiens),* but it also gave further support to the idea that the erect posture in the hominids developed earlier than the extension of the brain so characteristic of modern man. It further follows that the habitat of early hominids should not be sought in the primeval forest, but in open country. This realization is of the very greatest importance in the search for the geologically earliest hominids. We may deduce that the earliest hominids must have lived in the (later) Tertiary period. Such remains have recently been found in the Pliocene of South Asia and East Africa *(Ramapithecus = "Kenyapithecus")* (Fig. 47). In 1954, *Oreopithecus bambolii,* known since 1872 from late Miocene or early Pliocene lignites of Tuscany (Italy), was considered by the Swiss paleontologist J. Hürzeler as the geologically oldest hominid. Hürzeler reached this conclusion on the basis of a new examination of the remains of jaw and teeth and the fragment of an ulna then available.

This interpretation, which seemed inappropriate even on the basis of the morphological features, was further weakened by what could be deduced from the circumstances of the find regarding the environment in which this species had lived.

Experience has shown that vertebrates recovered from lignite, though not from its accompanying layers in the form of marls or sands, are either swamp or forest dwellers, not creatures of the open

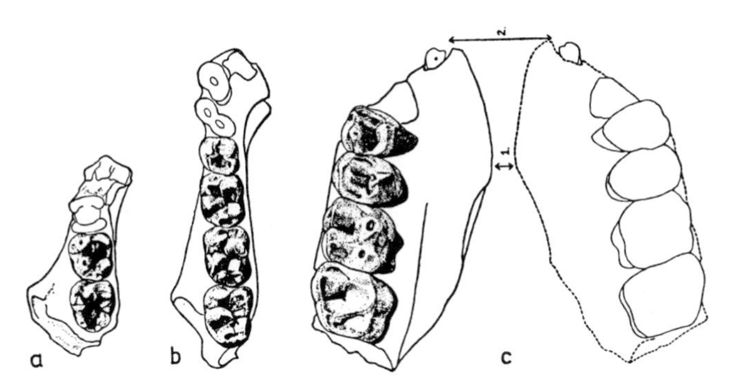

Fig. 47. *Ramapithecus brevirostris,* a hominid from the Early Pliocene of South Asia. a. Lower jaw with M_{2-3} sin.; c. upper jaw with P^3—m^2 dext. Dentition reconstructed by mirror image. b. *Dryopithecus (Sugrivapithecus),* a pongid from the Early Pliocene of South Asia. Notice short jaw and rounded curve of dentition in *Ramapithecus.* (After E. L. Simons, 1961 and 1963.)

country. In 1958, Hürzeler succeeded in recovering a fairly complete skeleton of *Oreopithecus bambolii,* which enabled an assessment to be made of the proportions of the limbs; this discovery fully justified the reservations mentioned above. *Oreopithecus bambolii* was definitely a climbing species that had long arms and resembled the modern apes, so it must have been an inhabitant of the primeval forest. It is unlikely that it had a bipedal manner of locomotion, such as is characteristic of hominids, not only because of the habitat and the extended arms but also because of the shape of the pelvis which, according to H. Schultz, most closely resembles that of a chimpanzee.

Careful investigations of remains of wood and leaves (by cuticle analysis), also fruits and seeds from the various layers of the Tertiary lignite have shown that various combinations of plants were involved in its formation, their sequence being presented schematically in Fig. 48.

Starting from an open lake in process of becoming dry land, something like the Everglades of Florida, a woody bog forms first, or flooded stands of swamp cypress (cf. the cypress swamps of Louisiana, etc.) consisting of *Taxodium, Glyptostrobus* and *Nyssa* (water tupelo),

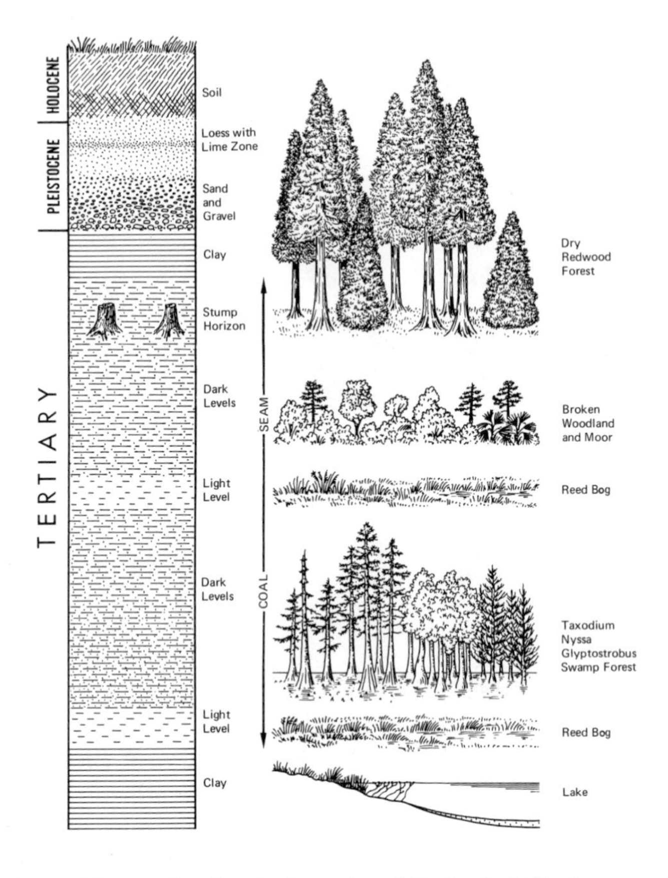

Fig. 48. Diagram showing the formation of lignite. Left, lignite seam with light and dark layers and a stump horizon; right, the plant communities from which it formed: reed bogs; swamp forests with *Taxodium-Nyssa-Glyptostrobus;* broken woodland and moor with alder, willows, conifers and dwarf palms; final stage—dry redwood forest. (After E. Thenius 1962, revised.)

which are finally replaced by a dry-standing forest of sequoia *(Sequoia langsdorffii,* a form related to the Californian redwood, *S. sempervirens);* this represented the climax or final stage and consequently persisted for longer. Hence, most of what are called the stump hori-

Fig. 49. Living "lignite forest" stages in the United States; a. reed bog, Everglades, Florida National Park; b. cypress swamps with *Taxodium distichum*, Florida National Park; c. climax stage, redwood forest, Sequoia National Park, California. Photos courtesy of USIS.

zons in lignite consist of sequoias. Conditions favored fossilization, however, only during the moist stages (Everglades, woody bog, and cypress swamps) so we cannot expect to find anything other than lake or swamp dwellers or forest forms. Thus, the remains of the inhabitants of steppe or savannah will only very rarely be found in lignite.

The following example shows how very much paleontology can be influenced by the gift of fossil preservation. The Recent African lung fish (Protopterus) is found in waters which dry up at times; during the dry period it buries itself in the mud and forms a wrapping round its body, consisting of mud outside and a cocoon of slime inside which hardens to the consistency of leather. During the dry-period sleep of the Protopterus, its vital functions are reduced to a minimum. The fish, because of its lungs, obtains the oxygen it requires through a breathing slit in the capsule until the next fall of rain restores the water; the fish then emerges from its softened capsule. In recent years, lung fish of this type (e. g., Gnathorhiza), which became fossilized during their dry-period sleep, have been found in the Pennsylvanian and Permian rocks of Texas. This shows that lung fish of the later Paleozoic era had already developed this peculiarity.

V. Fossils as Indicators of Time

Index Fossils as the Basis of Biostratigraphy

Every age has its characteristic fossils. This realization, first put to practical use by William Smith, an engineer who worked at the turn of the 18th to the 19th century, was developed into the concept of *index fossils* by Leopold von Buch. Index fossils enable the relative age of fossil-bearing sediments to be determined, which is a matter of great importance to the geologist, since a purely lithological (rock-based) analysis of the deposits is not adequate for determining their age. Breccias and conglomerates, sandstones and limestones, shales and marls, clays and oolites, and other inorganic sedimentary rocks can all, given the appropriate conditions, have formed at any period of prehistoric time. Only through index fossils which, of course, must be restricted to a particular zone, can sedimentary rocks be classified or placed in parallel according to age. This is the working area of *biostratigraphy,* also called biochronology (O. H. Schindewolf). Starting from the *law of superposition,*[1] first formulated in 1669 by Nicolaus Steno, comparative investigations have shown that there is always a regular sequence of plant and animal fossils in such rocks. This sequence is expressed in the time scale of Earth's history (see Table, p. 180), whose basic divisions were determined in Europe. These divisions were obtained from characteristic local profiles with an observable sequence of fossils. The perceptible change in the composition of the fossil flora and fauna — attempts were once made to explain this by worldwide catastrophes and subsequent new creations — is determined by evolutionary development. By this process the flora and fauna of the most recent geological periods (Tertiary and Pleistocene) have moved by small steps nearer to what they are today. This realization — which, however, is of only limited application — was originally invoked as the basis for the classification of Tertiary deposits, the percentage of Recent species found among the fossil fauna being regarded as the decisive factor.

Relative and Absolute Chronology (Geochronometry)

Evolutionary development, or phylogeny, is also a precondition for the presence of index fossils and hence for a relative chronology of the geological time as well. Fossils do not yield data on the absolute age of the layers in which they are found, but they do permit relative dating. An event normally resulting in the deposition of fossils is said to have happened not so and so many years ago, but in the Eocene age, Miocene age, etc.

Absolute dating, such as is obtained by *geochronometry,* is based on the one hand on the decline in the radioactivity of certain elements and on the other on the annual layering of many sediments. The latter led to the banded-clay or varve chronology, as it is called, which makes use of the principle that each year the water from the melting snows deposits a layer of clay in the lakes into which it drains. These banded clays date chiefly from the Late and Post-Glacial times of

[1] The law of superposition states that, with sedimentary rocks that have not been disturbed, the geologically oldest rocks are at the bottom and the geologically youngest at the top.

Northern Europe, when the inland ice which covered all of Scandinavia was slowly retreating northward. This is basically the same principle used in dendrochronology—comparative dating by means of annual growth rings in woody plants. While banded clays and annual rings have retained within their fabric a condition marking the passage of time on the basis of which it is possible to calculate age, radioactive elements in minerals conversely enable scientists to estimate how much time has passed since they were formed by interpreting the changes brought about by their constant rate of decay. They are "geological clocks" and may be compared to an hourglass: once we know how long the sand takes to trickle through, we can use the ratio between the amount of sand in the upper and lower compartments of the hourglass to calculate the time which has elapsed since the sand began to run. Radioactive elements occur in magmatic melts from which, as they cooled progressively, they were built into minerals formed by crystallization. The sand in the hourglass started to run at the point in time when the minerals were formed. The decay times of radioactive elements, or radioisotopes, because they are constant, form a basis for the absolute dating of periods in Earth's history. Every radioactive element has a characteristic half-life, the time it takes for half the element to decay, irrespective of such physical factors as temperature or pressure. This half-life has been determined experimentally and is an absolute constant. For uranium it is about 4.5 billion years, for thorium about 15 billion years. Half-lives are true atomic clocks. Such differences as there are in the data obtained from them are generally due to sources of error in the correlation of magmatic rocks, which contain radioactive elements in their minerals, with fossil-bearing deposits.

The most important radioactive elements are uranium (atomic weight 238), actino-uranium (235), and thorium (232). The decay of these elements, which consists in the emission of α- and β-particles and also γ-radiation, gives rise in due sequence to various other elements, only the last in the sequence being stable; this last element has the properties of lead and has an atomic weight of 206, 207, or 208, respectively, according to its original element.

Radioisotope dating methods are of fairly recent development. The best-known is radiocarbon dating, which makes use of ^{14}C. Normal carbon has the atomic weight 12 and is thus written ^{12}C. The variants with atomic weights 13 (^{13}C) and 14 (^{14}C) are much less common. ^{14}C is present in the Earth's atmosphere in the form of carbon dioxide and is taken up by plant organisms via assimilation. By this route ^{14}C either returns to the atmosphere via the natural cycle or is deposited in the organic residue in the soil, where the decay which began upon the death of the organism continues. Since the half-life of ^{14}C is short (5,730 years) ,^{14}C dating is applicable only to the last 50,000 years. It has provided the paleontologist with valuable information about the point in time when Pleistocene mammals became extinct.

Another radioisotopic method is the potassium—argon method ($^{40}K \rightarrow {}^{40}Ar$). Its half-life makes it specially valuable for determining sequences in the Tertiary period, which is not susceptible to fine distinctions by the uranium method. The starting materials are biotite, glauconite, muscovite, and hornblende which, however, all react differently in their argon loss at high temperature and may thus yield conflicting values for geological age. The rubidium—strontium method is also important. Usually, only the geologically most recent event associated with a rise in temperature can be dated by these methods.

Taken as a whole, this complex of "geological clocks" permits absolute dating not only of the Phanerozoic eras (the Paleo-, Meso-, and Cenozoic) but also of the Precambrian era. From these clocks it is thus possible to draw conclusions about the probable age of the Earth, which has recently been computed as at least 4.5–5 billion years. This time span is about seven to eight times as long as the time which has elapsed since the beginning of the Phanerozoic era, which was when fossilization really started. Stated in terms of the distance from New Haven to Chicago, this would mean that you would not find much in the way of living organisms until you reached South Bend.

For the paleontologist, however, what is important is the absolute dating of those time spans which have bequeathed us rich fossil-bearing deposits, beginning with the Cambrian period, some 550 million years ago. On the strength of these absolute dates, it is possible to make estimates of the rate of evolution, that is, of the speed with which the various species have developed; this is a subject to which we shall return in Chapter VI.

The fluorine test, as it is called, must also be mentioned, as it can be a great help to the paleontologist. It determines the fluoride content of bones and teeth, which usually increases with increasing geological age. As this increase, unlike the decay of radioactive elements, is strongly influenced by various external factors, the fluorine test is of limited application in determining age. It has been most helpful so far in identifying fossil frauds and in evaluating the origin of fossil remains subjected to transport.

The classification of sedimentary rocks by means of fossils has mostly been possible where there was a rich fauna present and the original deposition has not been disturbed, as in the South German Jura Mountains. Here the fundamental research on ammonites carried out by F. A. Quenstedt (1809–1889) and his pupil, A. Oppel (1831–1865), formed the basis for the classification of the Jurassic period, which has become a classic of its kind, although preceded by similar classificatory investigations in Great Britain. It is logical that stratigraphic classification should have had its origin in regions where there was little or no tectonic disturbance and not in rock series subjected to strong tectonic stresses, such as are found in the Alps or the Himalayas.

Other regions which have become classic because they were used in establishing the chronology of a period are the Welsh mountains (Cambrian and Ordovician), the Baltic (Silurian), Germany and Belgium (Devonian), Belgium (Carboniferous), Central Germany (Permian and Triassic), Southwest France and West Switzerland (Cretaceous), and the Paris, London, and Vienna Basins (Tertiary). From these, too, are derived some of the stratigraphic names based on geographic terms: Cambrian from Cambria, the old name for Wales; Ordovician from the Ordovicii tribe of N. Wales; Silurian from the Celtic tribe of Silurii in S. Wales; Devonian from the English county of Devonshire; Permian from the district in the Urals; and Jurassic from the Jura Mountains of S. Germany. Thus, the stratigraphic system was set up within a relatively short period, roughly between 1820 and 1850, and its basic features are still the same today.

Index fossils include principally those groups of organisms which show the greatest changes within a particular time span and which are sufficiently common and widely distributed. The most important of them are: trilobites (Cambrian-Silurian); graptolites (Ordovician and Silurian); ammonites (Devonian-Carboniferous, Triassic-Cretaceous); nautiloids (Early Paleozoic); belemnites (Jurassic-Cretaceous); Foraminifera (Permocarboniferous, Cretaceous, Tertiary); brachiopods (Paleozoic-Mesozoic); echinoderms (Paleozoic-Tertiary); reptiles (Mesozoic) and mammals (Tertiary-Pleistocene) (see Fig. 50). It is obvious that the remains of the index fossils must be sufficiently characteristic to enable them to be distinguished without difficulty from individuals of similar species and to enable clear distinctions to be made between species.

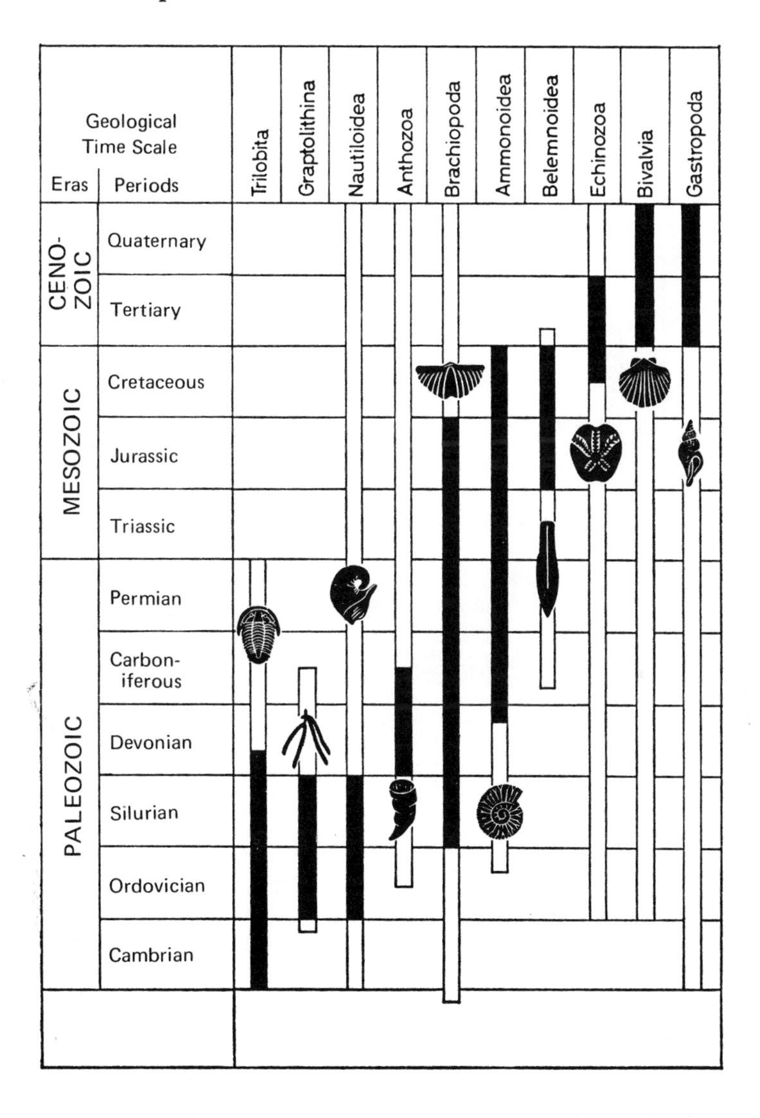

Fig. 50. Time scale showing the most important groups of macrofossils among the invertebrates. White column, time span of distribution; black column, time span in which they are important as index fossils. (After E. Thenius, 1970, revised.)

Ortho- and Parachronology—the Concept of Zones

The fundamental classification as determined by the index fossils (e. g., ammonites, trilobites, graptolites, conodonts, and ostracods) is the so-called orthostratigraphy or "orthochronology," in contrast to the parastratigraphy, or "parachronology," which is based on an ecologically conditioned change in the fauna or flora.

The basic unit of time which can be delineated by means of its fossils is the *zone,* also called biozone. Its duration is determined by the life of a species (A. Oppel). It therefore follows that ideal index

General Outline			Level Make-Up		Zone Fossils
Lias	Sinemurium	β_2	Lower Clay Beds	Oxynoticerata Levels	Echioceras raricostatoides 
					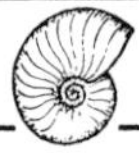 Oxynoticeras oxynotum
		β_1			Asteroceras obtusum 
					 Asteroceras turneri
		α_3	Gryphaea Limestone	Arietita Levels	Arnioceras geometricum 
					 Vermiceras spiratissimum
	Hettangium	α_2	Angulata Sandstone	Schlotheimia Levels	Schlotheimia angulata 
		α_1	Psilocerata Clays	Psilocerata Levels	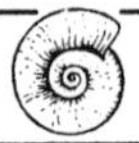 Psiloceras planorbe

Fig. 51. Ammonites as zone fossils in the Lower Jurassic Lias of Swabia, South Germany. (After E. Thenius, 1970, redrawn.)

fossils are the various members of an ancestral series which differ specifically from one another because each species is replaced by the next younger one, thus leaving one characteristic species for a particular zone (Fig. 51). Larger chronological units are called stages, epochs, and periods. In German, the word "Formation" means the same as period, but in the English-language literature it is used in quite a different sense to mean a more or less uniform rock complex occurring characteristically in a certain terrain.

Intercontinental Correlation

The definition of the various epochs and higher chronological units is a matter for international agreement. There are a number of permanent stratigraphic committees of experts whose task is to fix the definition of such terms in an internationally acceptable manner and to settle controversial matters. For example, the so-called Danian, which is intermediate between the Cretaceous and the Tertiary, is regarded by some authors as the latest stage of the Cretaceous period, and hence of the Mesozoic era, and by others as the earliest stage of the Tertiary period, hence of the Cenozoic era. Such differences of opinion usually depend upon differences in the interpretation of the time of extinction or of the first appearance of index fossils, where this appearance was not simultaneous everywhere. Apart from such differences, it has become clear that the phylogenic development of the various groups of organisms definitely does not proceed uniformly and hence that sections of flora and fauna need not always occur together (see Geological Time Scale), although the former are always earlier.

Realizations like these, plus the fact that marine transgressions onto land areas have obviously occurred time and again, have induced geologists to try to draw the lines dividing stratigraphic units according to tectonic phases, transgressions, and other inorganic phenomena or their consequences (e. g., transgression conglomerates). Here it must be stressed that inorganic phenomena are reproducible, whereas those resting upon phylogenetic development are unique and irreversible. Quite apart from this, a reliable worldwide parallel is possible on the basis of fossils alone, although this is properly part of stratigraphy. To be sure, even when fossils are used, there are often great difficulties in the way of equating or parallelizing fossil-bearing deposits of regions which are geographically distant from one another, and particularly in making intercontinental correlations. Pelagic organisms i. e., those swimming or floating in the high seas, have been the most suitable for this task, since their distribution is worldwide, at least within certain climatic zones. These include on the one hand planktonic organisms (minute, passively floating aquatic organisms, such as Foraminifera, diatoms, silicoflagellates, and coccoliths, including discoasterids) and on the other hand nektonic organisms (free-swimming aquatic animals, sometimes of remarkable size, like sharks, bony fishes, whales, and cephalopods).

The planktonic organisms have acquired special importance because they are so microscopic that their skeletal remains may be contained in large numbers even in very small amounts of material, e. g., cores obtained by boring. The globigerinids, in particular, play a leading role in the intercontinental correlation of Tertiary deposits. Very recently what are known as nannoplankton (dwarf plankton under 40 μ in size, and hence considerably smaller than normal plankton; mainly discoasterids and calcareous flagellates) have been used with great success for the correlation of Tertiary deposits.

Observations of their "speed of migration" enable the time span needed for a worldwide distribution of marine organisms (and continental ones, too) to be ignored because the several hundred years required (up to a maximum of several thousand years) cannot be detected by our present methods of measuring geological time. Apart from active expansion by nekton, an important role in the worldwide distribution of planktonic organisms is played by ocean currents, which influence both the larval and the adult stages.

Biostratigraphy and Ecology

In contrast to these universally distributed marine planktonic and nektonic organisms, the dwellers on the ocean floor, or benthic organisms, usually have a spatially very restricted distribution. They often occur only in particular marine basins (e. g., the Paris, London, and Belgian Basins) and are not found in neighboring areas where they are replaced by other species and subspecies. There are thus true geographically determined differences in contemporary flora and fauna which enormously complicate correlation between deposits of the same age. An additional factor is facies differences, which alone can make it impossible to compare even neighboring deposits (cf. Chapter IV, p. 57). So the ask of biostratigraphy is to draw a distinction between differences determined by facies and those due to age. The correlation of deposits within a particular sedimentation region (sediments of margin and basin facies, such as the Leitha limestones and the Baden Tegel in the Miocene of the Vienna Basin) may be effected by means of facies-breaking index fossils, i. e., organisms not bound to a particular facies, which can occur (to continue with the example quoted above) in sand, limestone, and clay sediments. Here palynology can be very helpful — the pollen of wind-pollinated plants is carried extremely long distances by the wind and thus comes to rest in a wide variety of sediments. Their contemporaneity can be established by the agreement of the pollen spectrum (Fig. 52). Thus, pollen and spores play an important part in the correlation of terrestrial with marine sediments.

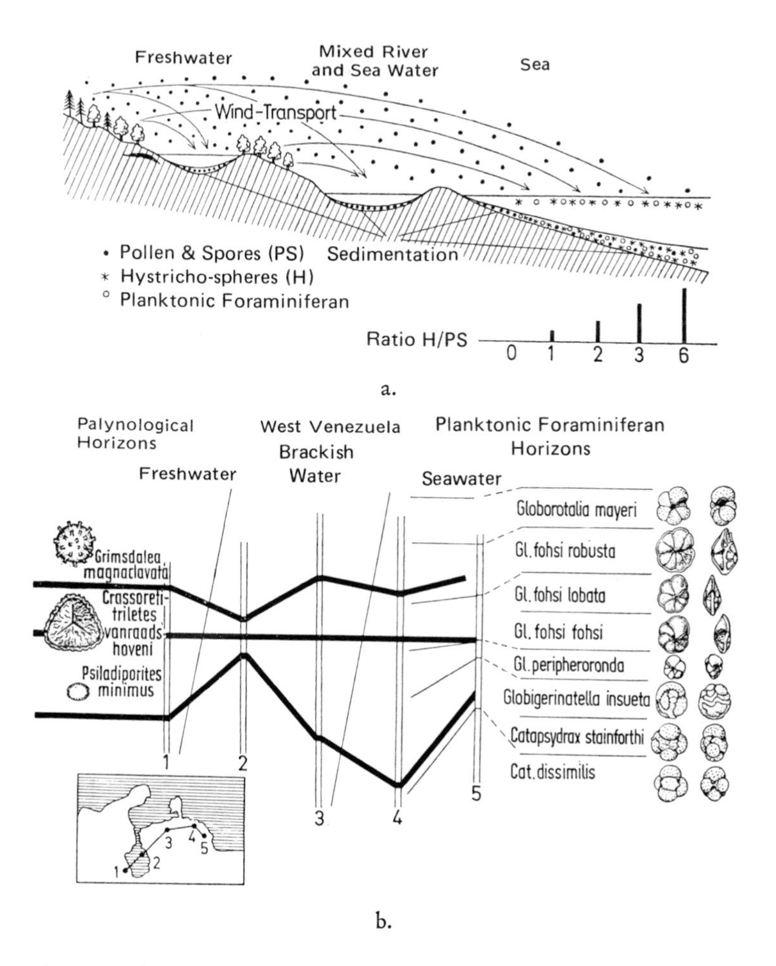

Fig. 52. Palynology and stratigraphy. Parallelization of non-marine with marine deposits by means of widely dispersed pollen and spores. Above, diagram (after G. Charrier, 1965, redrawn and revised); below, correlation of five bore holes of Miocene freshwater, brackish-water and marine deposits. Right, planktonic foraminiferan horizons. (After C. A. Hopping, 1967, revised and redrawn.)

Further difficulties in determining the chronological sequence arise from the fact that index fossils may be redeposited in geologically younger deposits. Redeposition can occur with reworking in coastal areas through wave action, or by submarine slides or so-called turbidity currents. Obviously fossils in heterochronous allochthonous deposits (see p. 28) cannot be used for age determination. Generally, though not always, fossils that have been redeposited can be recognized by their state of preservation.

On the other hand, the exact specific or subspecific determination of index fossils is a necessary precondition for a reliable stratigraphic interpretation. The results of convergence or parallelism must be excluded without any shadow of doubt. The characteristic features of the various groups of index fossils are so varied that it is impossible to discuss them within the scope of this book. Let us just mention the following features which are important for systematic classification: for ammonites, the form of the suture lines; for (Paleozoic) Nautiloidea, the overall shape, structure, and position of the siphon; for corals, the structure and arrangement of the septa; for lamellibranchs, the overall shape and development of the valve and the shell sculpture; for brachiopods, the shape of the shell, the pedicle foramen, and internal loop; for large Foraminifera, the chambers and type of the spirals; for mammals, the dentition.

Nevertheless, the fact that it is not absolutely necessary to have a good knowledge of the systematic position of index fossils is witnessed by the sporae dispersae (pollen grains and spores which are found in isolation); aptychi (opercula of ammonites) and otoliths (calcareous concretions in the ears of fishes). The assignment of such easily preserved remains to particular species is almost always doubtful. However, all that is really required in index fossils is that the various "species" should be distinguishable from each other, thus permitting identification.

For the stratigraphic classification of marine deposits such traditional index fossils as ammonites, graptolites, belemnites, brachiopods, and corals are usually relied upon, although in recent years microfossils (chiefly Foraminifera) have become more important as they become more widely used; however, ostracods (crustacea in a two-valved carapace) and plant fossils play no significant part. For the classification of nonmarine deposits, in addition to vertebrates and nonmarine molluscs, ostracods and plant remains are the most important index fossils. The latter, as macrofossils, are indispensable for the division and correlation of the Carboniferous coalfields or seams, whereas for the classification of the mainly Tertiary lignites, plant microfossils (pollen and spores) are more important than macrofossils (remains of wood and leaves, fruits and seeds).

Palynology and Stratigraphy

Pollen and spores, particularly in very recent times, have played a decisive role in determining the relative age of salt deposits; for in salt and its accompanying clays (so-called Haselgebirge) of the alpine

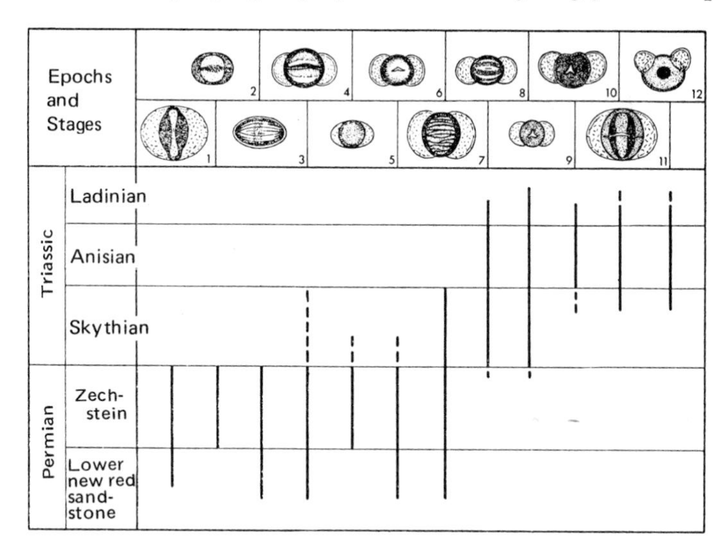

Fig. 53. Spore stratigraphy. Temporal distribution of airborne spores from the Permian and Triassic. 1—7, typical spores from Alpine halites which characterize the dating as Upper Permian (Zechstein). 1. *Paravesicaspora splendens;* 2. *Lueckisporites parvus;* 3. *Vittatina ovalis;* 4. *Lueckisporites microgranulatus;* 5. *Klausipollenites schaubergeri;* 6. *Jugasporites perspicuus;* 7. *Striatites jacobii;* 8. *Taeniaesporites novimundi;* 9. *Triadispora staplini;* 10. *Tr. falcata;* 11. *Illinites kosankei;* 12. *Microcachrydites fastidioides.* (After W. Klaus, 1965, from E. Thenius, 1970.)

salt domes (e.g., in Austria, Salzkammergut, Hall in Tirol), other fossils are either not found at all, or very rarely. Only the highly differentiated pollen spectra offer a sequence going back to the Late Permian period; thus they could be correlated with the Zechstein salts of the Rhine valley in Germany, etc., once a general sequence had been derived going back to the Early Triassic period (Skythian stage or Buntsandstein, Fig. 53).

The initial importance of pollen analysis, however, lay in the investigation of the Late and Post-Glacial forest history of Central Europe, which was reconstructed on the basis of large numbers of pollen diagrams obtained from peat profiles. The gradual reafforestation of areas that were bare of forest in the Ice Age also allows a chronological reconstruction of Late and Post-Glacial times. Quite recent investigations have even enabled the various warm periods within the Ice Age to be identified and distinguished on the basis of the percentage composition or the slightly deviant sequence of the appearance of the various species of trees. As should have become clear from the foregoing, the classification of Quaternary deposits aided by pollen spectra (which are pooled to give a pollen diagram) depends not on the index fossil principle but upon ecologically determined processes (reofforestation or retreat of tree vegetation),[2] whereas the composition of pollen flora at the beginning of the Ice Age — apart from the occurrence of several relicts from the Tertiary period — is not essentially different from that found today.

In conclusion, let us say a little more about the significance of trace fossils as time markers. It is very exceptional for trace fossils to be assignable to a particular time, although it is the case for the excrement trail *(Tomaculum problematicum)*, which so far has only been found in the Ordovician of Bohemia, the Rhineland, Thuringia, Britain, and the South of France. In general, the stratigraphic usefulness of trace fossils runs up against difficulties, as they are often connected with facies (e. g., flysch trace fossils: helminthoids, chondrites, and fucoids in Cretaceous flysch, which has a predominantly marl-type development; snail traces, *Taonurus* and *Palaeodictyon* in the Eocene flysch, which consist mainly of sandstones) and are frequently far less differentiated than the animals which produced them (e. g., movement traces — such as tracks and burrows). For this reason a variety of organisms have to be considered as possible producers of trace fossils of a particular type [e. g., star traces, which may be left by shrimplike crustaceans *(Corophium)* or clams *(Scrobicularia)*].

[2] In Quaternary pollen diagrams, only tree pollens (TP) are divided according to genera, while the mass of nontree pollens (NTP) are usually presented as an entity.

VI. Fossils and Evolution

Paleontology as a Chief Support in the Study of Evolution

As mentioned in the Introduction, research into the origin and development of species has become one of the main tasks of paleontology. Only through fossils can the actual course of phylogeny be charted, a fact recognized today by all serious biologists, although sufficient proof of evolution is supplied by the hierarchic multiplicity of forms of organisms alone. However, it does not get us very far simply to say that there has been an evolutionary development. The views of various research workers on phylogenetic relationships, usually expressed in a "phylogenetic tree," frequently differ quite widely. On the grounds of their two-dimensional presentation alone, such phylogenetic trees can, of course, only be regarded as schematic; yet the differences of opinion about the way in which species have evolved usually depend more on divergent interpretations of individual features, or their homology. The concept of homology is basic to any taxonomic or phylogenetic investigation. Organs that are homologous have a similar origin. For instance, the wings of a bird and the fins of a whale are homologous, but the wings of birds and the wings of insects are not. The most important criterion of homology is the positional relationship in the organ in question. Despite vast differences in shape, usually determined by function, the positional relationship is capable of distinguishing organs that are homologous.

Gaps in the Fossil Record

The importance of paleontology was recognized rather late in the day, and the reason for this is to some extent historical. Even Cuvier, the founder of vertebrate paleontology, asserted the constancy of species and maintained this point of view against those of his contemporaries who disputed it — specifically, Lamarck and Geoffroy St. Hilaire, who were persuaded that organisms had indeed enjoyed an evolutionary development. The assumption of the immutability of species led to the Catastrophe Theory, which found its most extreme protagonist in the French paleontologist A. d'Orbigny. In accordance with his theory, he ascribed the differences in fossil flora and fauna in different deposits of different ages, which were well-known even in his day, to reiterated acts of creation (cf. p. 7). Not until Ernst Haeckel (1834–1919) carried out his pioneering work inspired by Darwin's fundamental book, *The Origin of Species*, published in 1859, did paleontology receive the correct orientation so that it could be used to explain the evolutionary phenomena which had taken place in the course of Earth's history. The Swiss, L. Ruetimeyer, was one of the first to come seriously to grips with the problems which Darwin's book posed for paleontology. In 1856, however, the Viennese paleontologist M. Hörnes had already noted phylogenetic changes in fossil snails of the genus *Cancellaria.*

Nevertheless, at that time paleontology was unable to play any significant part in the evaluation of evolutionary questions. On the one hand, there was the well-worn phrase of the zoologists about the "gaps in the fossil record"; on the other, doubts were voiced on all sides concerning the capacity of fossils to furnish proof in evolutionary matters. This atmosphere greatly inhibited progress. There was also the fact that the views represented by paleontologists on the mechanism of evolution could not by any stretch of imagination be reconciled with the findings of genetics.

What is the status, as we see it today, of the "gaps in the fossil record" and the "proofs" supplied by paleontologists in evolutionary matters? In Darwin's day, indeed, such proofs in the shape of transitional forms were almost entirely lacking, which was what gave rise to Darwin's expression "the missing link." However, we should not overlook the fact that ideas about such "missing links" or, as we should now prefer to call them, "connecting links" have undergone a change with the passage of time, particularly since the *mosaic modus of evolution* (or Watson's law, as it is also called) has been recognized (see p. 96). When the "transitional forms" were available only as scattered specimens, they very seldom agreed with the claims laid down by theoretical considerations. Not until many more fossils had been discovered, thus providing an insight into the enormous variety of forms which had existed in earlier times and the composition of the countless populations which had followed one another in temporal succession, did a fundamental change come about in ideas on the subject, but this change has only been during the last several decades. It became simultaneously apparent that the findings of paleontology do in fact agree with those of genetics. Consequently, paleontology has become one of the most valuable props of evolutionary theory. This about-face is due in no small measure to the somewhat faster rate at which fossil populations have been recovered in recent times,

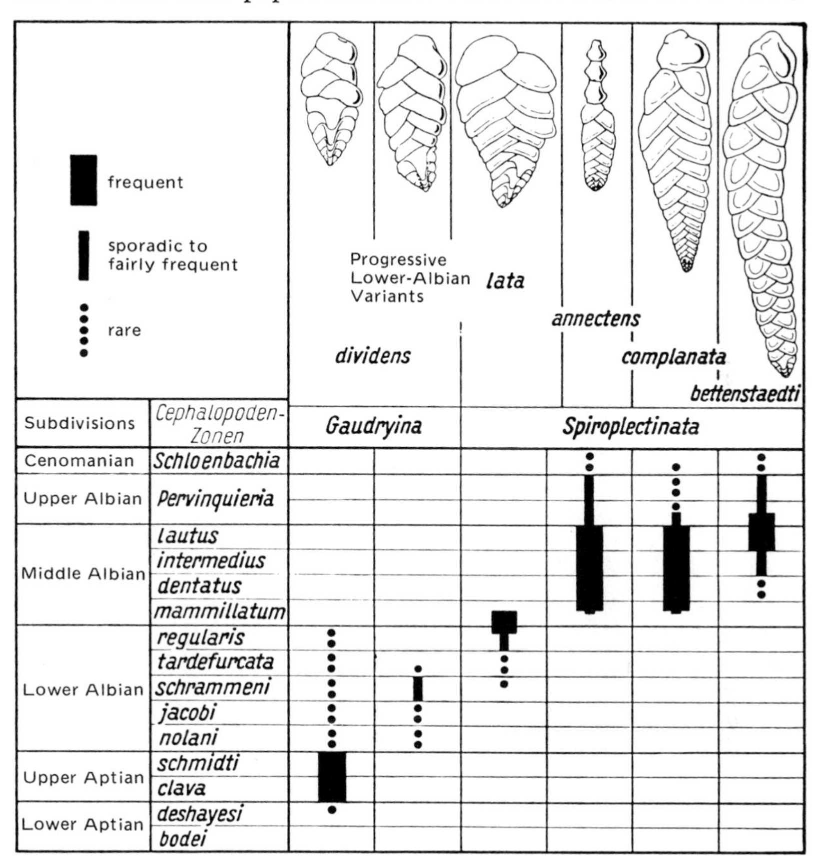

Fig. 54. Vertical range and frequency of foraminifera *Gaudryina dividens* and the four species of *Spiroplectinata* from the Cretaceous of Northwest Germany exemplifying species and genus succession. (After B. Grabert, 1959.)

particularly in micropaleontology. Thus, this branch of science, which was initially developed for the benefit of field geologists, has become extremely significant in evolutionary matters. It is perfectly possible to make population studies of micromammalia from fissure fillings or limnal sediments from which they can be extracted by elutriation.

This substitution of population studies for studies of single individuals, made possible chiefly by microfossils, has not only provided much interesting information but has also gone far to invalidate the old reproach of "gaps in the fossil record" — at least, in this sector. The immense importance of these fossil populations is that they are derived from profiles which often represent a considerable thickness of sediment — for instance, in deep bores. Fossil populations of this kind, and macrofossils, too, collected through the depth of the profile and interpreted accordingly, enable us to determine how gradual

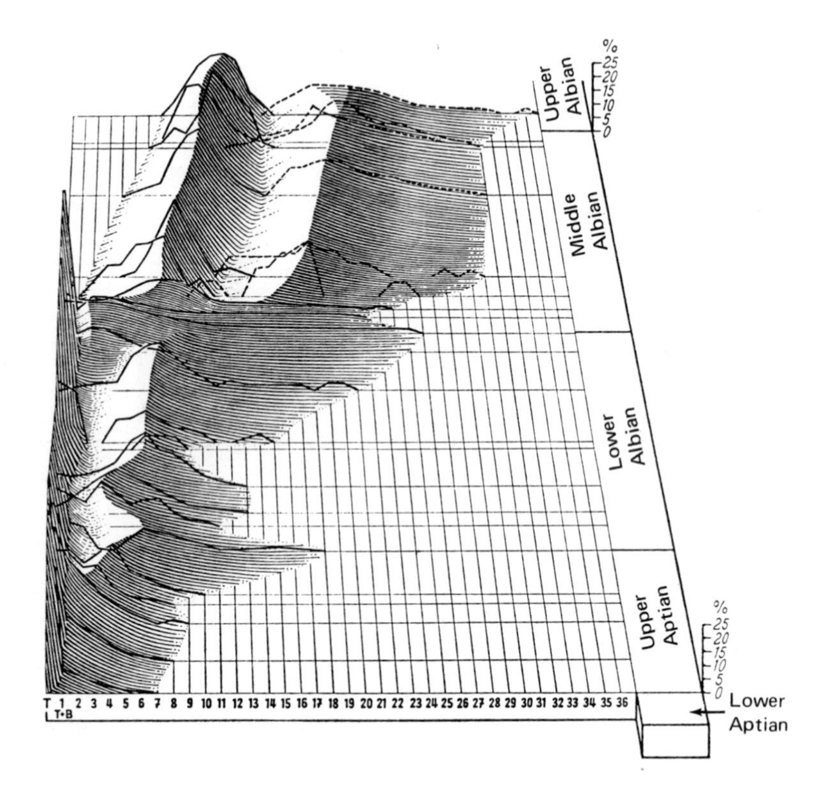

Fig. 55. Three-dimensional diagram showing the variation curves in species changes due to speciation of the foraminifera shown in Fig. 54 over a time span of several million years. Left the *Gaudryina* lineage appears as a steep, rocky ridge. From this a side chain splits off representing the progressive *Gaudryina dividens* variants. The valley and side chain merge into a broad downward slope that stands for the great range of variation of *Spiroplectinata lata,* from which proceed by differentiation the two branches of *Sp. annectens* and *Sp. complanata.* (After B. Grabert, 1959.)

displacements occur in characters within certain populations in the various sequential layers, with the result that the final population is totally different from the initial one. Figs. 54 and 55 show one example where not only the subspecific and specific differentiation during phylogeny can be observed, but even the generic, although the last is not acceptable, because shells of Foraminifera are not really suitable for studies of this kind. Zoologists today have neither enough time nor enough sequences of generations to establish generic differences; the most they can hope to do is to differentiate subspecific differences, perhaps in the form of geographical races.

Problems of Evolution

The examples cited above indicate how subspecific and specific changes may be verified by fossil material as the summation of minute evolutionary steps over a long period of time. The question then arises as to whether it is possible to explain the *whole* process of the evolution of organisms by such tiny evolutionary steps, i. e., by what is called microevolution or microphylogeny, or whether there is another set of rules which apply to transspecific evolution, hence macroevolution or macrophylogeny — what G. G. Simpson called mega-evolution. Is it necessary to have "great leaps forward," creating quite new types of organisms, and can these new types be demonstrated, or is it possible to explain everything by microevolution? This controversial area is very closely associated with the aforementioned "gaps in the fossil record," but it can only be treated adequately in combination with the findings of genetics.

As far as fossil finds are concerned, where unbroken profiles exist with a rich content of fossils, only the gradual type of changes referred to above have been detected. What had seemed to be larger jumps, with corresponding transspecific deviations, are quite simply — apart from ecologically determined changes — gaps created by erosion. This fact has been shown by, among others, R. Brinkmann who conducted statistical-phylogenetic studies on ammonites *(Kosmoceras)* from the English Jurassic (Dogger) and R. Kaufmann who studied trilobites *(Olenus)* from the Swedish Cambrian (Fig. 56). These results agree with the experimental results of geneticists, who have found only very small jumps which — when they are heritable — are found to have been caused by mutations (changes in the genes).[1] Simultaneous complex mutations, postulated as necessary for the sudden emergences of completely new types of organization, have not so far been reliably found to occur. Genes mutate randomly and independently, so that the probability of a viable organism originating from a complex mutation — involving, say, five genes at once — is very remote. In fact, even under the most favorable conditions (a population of

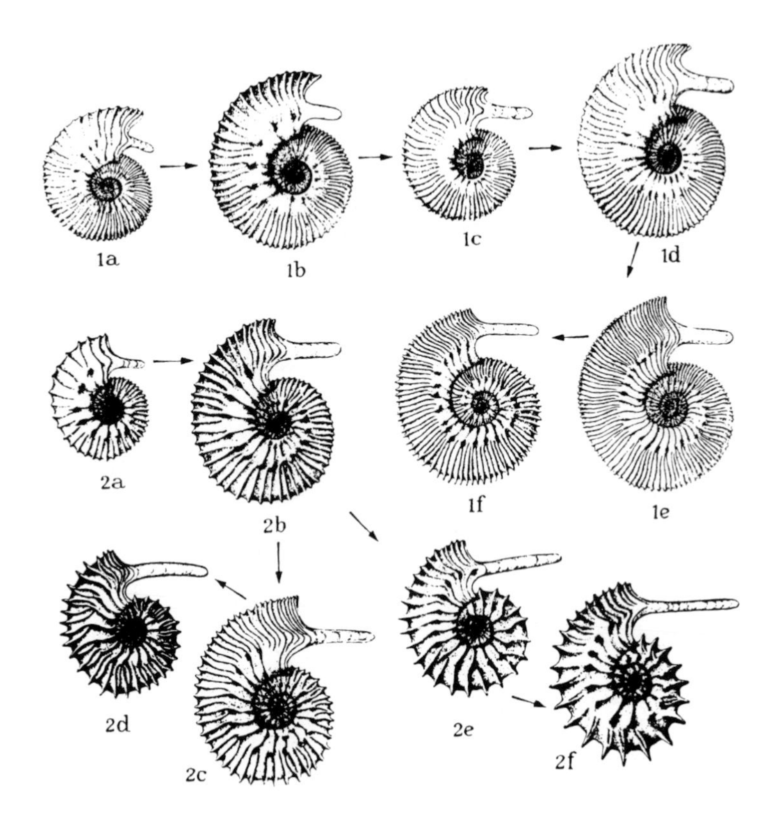

Fig. 56. Evolution of the cosmoceratids (ammonites) from the Middle Jurassic of Peterborough, England. Note development of ornamentation and ears. 1 a—f, subgenus *Gulielmiceras;* 2 a—f, subgenus *Spinikosmokeras.* (After R. C. Moore, 1957.)

[1] The phenomenon of polyploidy (multiplication of the gene stock: diploidy, tetraploidy, and so on) is known to occur mainly in plants, where it can produce very vigorous individuals; it is important ecologically because it results in competitive ecotypes, but it is not significant for evolution as a whole.

10^8 individuals with a generation span of a single day and a mutation rate of 0.00001), the calculation of probability shows that this formation would only be likely to happen once in the course of a period of time exceeding the age of the Earth by 100 times (G. G. Simpson). Consequently, the findings of genetics cannot be brought into line with the possibility of "great leaps forward" involving extensive structural changes. In the opinion of the proponents of the above theory, such structural changes could only have been achieved in a plastic, early juvenile phase of individual development, which would explain why no mature forms have ever been recovered as fossils.

This brings us back once again to the gaps in the fossil record. If we consider the enormous element of chance in the embedding and fossilization of organic remains, and the further element of chance in their being found at all and then reaching the hands of those qualified to interpret them, we need not be in the least surprised that there should be gaps in the paleontological record, at any rate where macrofossils are concerned. A distinction can be made between permanent gaps in our knowledge and provisional ones which can, in principle, be bridged. The reason for the permanent ones is the limited capacity of soft tissues to be preserved. With very few exceptions (see p. 18) the whole of anatomy as it applies to the soft tissues is a closed book to the paleontologist. The single exception concerns the anatomy of the brain of higher vertebrates where, as described on p. 13, the formation of natural casts provides an acceptable substitute. The provisional gaps in knowledge are growing smaller, as the following data will show. In 1859, when Darwin published his *Origin of Species,* no fossils had been discovered which could have supplied information on the origin of man[2] or the origin of the mammals. Since then, not only have numerous Late Pleistocene Neanderthal finds been made in Europe, Asia Minor, and North Africa, but there is also a whole range of material from the Middle Age "Pithecanthropus" group and the Lower Pleistocene australopithecines, material which is constantly being added to. These finds have provided important data and new knowledge which would have been beyond the wildest dreams of paleontologists even 30 years ago. Quite recently, moreover, there

[2] The classic discovery of Neanderthal man was, indeed, made in 1856 by a schoolmaster, J. C. Fuhlrott, but at that time, mainly owing to the authority of the opinion of the anatomist, R. Virchow, its significance was not appreciated—except by Fuhlrott himself.

have been finds from the Tertiary Period which can probably be classed as hominids.

Because of fossils, we also now know much more about the origin of mammals. There is no longer any difficulty in understanding how mammals evolved from reptiles — on the contrary, the amount of material available is so extensive that it has been necessary to create an entirely artificial boundary between reptiles and mammals. This will probably surprise the nonexpert who supposes there to be fundamental morphological, anatomical, and physiological differences between reptiles and mammals. This is true enough of Recent members, but it is not true of the therapsids — the ancestral group to the mammals within the order Reptilia — and the geologically oldest mammals. Several phylogenetic lines can be distinguished among the therapsids of the Triassic period for example, the ictidosaurs, cynodonts, and tritylodonts, which all show the "trend" toward mammal-like reptilian forms (Fig. 57). The gradual changes in shape affect not just the skull and the rest of the skeleton, but also the dentition and, as may be indirectly inferred, physiological characters, too, and other features not susceptible to fossil preservation. These mammal-like reptiles were undoubtedly warm-blooded, had a hairy covering, and probably suckled their young from milk glands. The decisive characters for the paleontologist, however, are the structure of the lower jaw and auditory region, which can readily be assessed from fossils. In mammals, three types of auditory ossicles are developed in the middle ear — malleus, incus, and stapes — while reptiles never had more than one, the stapes. In the nineteenth century the anatomist C. Reichert had already postulated, on the basis of investigations of mammalian and reptilian embryos, the homology of the malleus with the articulare, a bone of the lower jaw, and of the incus with the quadratum, a bone of the skull. Fossil evidence has fully confirmed this theory, now called the Reichert-Gaupp theory in honor of the great services rendered by E. Gaupp in proving the existence of genuinely double-jointed animals (e.g., *Diarthrognathus* and *Probainognathus*). At the same time the apparently unbridgeable gap between reptiles with the so-called primary lower-jaw joint between the articulare and the quadratum and mammals with the secondary lower-jaw joint between the dentale and the squamosum, has been filled in. Here, too, the fossil record shows that small, gradual changes and not great leaps brought about these differences. Further, fossils

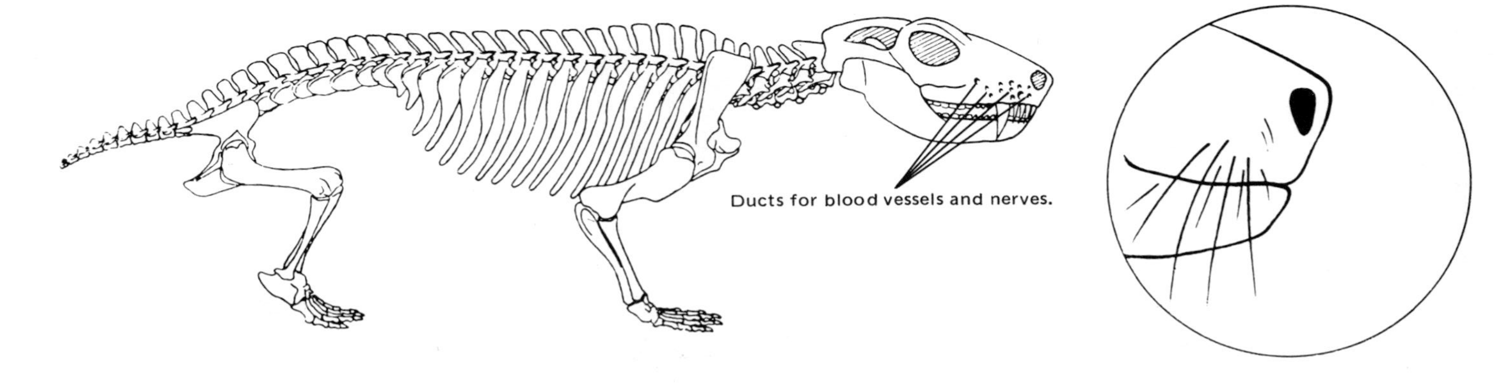

Fig. 57. Mammal-like reptile (therapsid) from the Triassic of South Africa. Reconstruction of skeleton. Note ducts or pits for nerves, blood vessels and hair roots in the upper muzzle, suggesting formation of lips and whiskers (see inset). This leads indirectly to the assumption that the animal had a coat of hair to protect it and was warm-blooded.

teach us that the several characteristics within the various lineages did not evolve synchronously, that is, at the same speed, and that, for this reason, some therapsids never evolved into mammals. The regularity with which this was found to be so justified the establishment of the law of evolution describing the so-called mosaic mode of evolution, named after the British vertebrate paleontologist D. M. S. Watson.

The examples cited above have also touched on the problem of the cause of phylogenic development which in turn posed the question of the law governing the processes of evolution. This is a complex of questions which can only be dealt with marginally, since they fall within the province of the geneticist rather than that of the paleontologist.

The raw material of evolution is that an organism should have an *excess of offspring* and *mutability,* as was correctly noted by Darwin. Since hereditary changes, or mutations, occur in a random fashion and yet evolution is emphatically not a random process, there must be a factor which applies direction to it in free nature. This factor Darwin called *natural selection.* This is the mechanism which brings about gradual changes with the passage of time until full adaptation is finally achieved. Most biologists are convinced that the mechanism of selection is the *only* factor guiding evolution.

Increasing improvements in adaptation have been adduced in the past by paleontologists as "proof" of orthogenesis or orthoevolution, by which they mean a directed development, and used to counter the arguments of geneticists. The best-known example of "orthoevolution" is the "genealogy of the horse," an expression coined by the American paleontologist O. Marsh. Quite apart from the fact that it should be called orthoselection rather than orthoevolution, the phylogeny of the horse, as was shown by G. G. Simpson, who was at that time the leading mammalian paleontologist, is certainly not a good example of straightforward development. There are many lineages which have to be distinguished within the family of the equids, and only one of these is still extant. The one-toed horses of the present are merely a sideshoot; the main lineage, Anchitherians, became extinct as long ago as the Later Tertiary period (Fig. 58). The evolution of the horse, attested by fossils dating from the Eocene age, has also shown that, if false interpretations from a phylogenetic point of view are to be avoided, it is most important to distinguish between se-

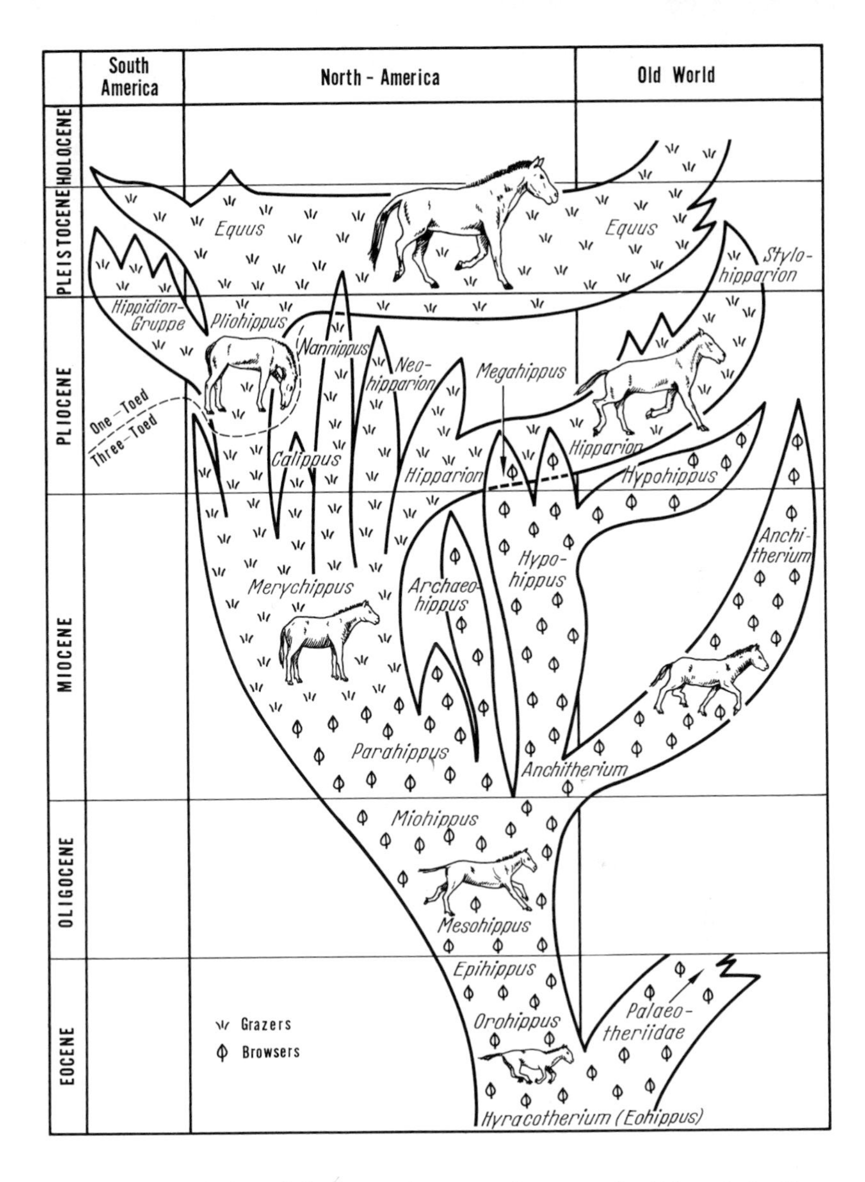

Fig. 58. Evolution of the horses. Reconstructions to scale. (After G. G. Simpson, 1951, revised.)

quences of changes and sequences of ancestors. Thus, in Europe the following confirmed genera of equids form a single sequence of changes: *Anchitherium* (Miocene), *Hipparion* (Pliocene), and *Equus*

(Quaternary).[3] All these came from North America to Europe over what was then a land bridge at the Bering Straits. Furthermore, the relationships with the environment can be seen not only in food and hence dentition (Fig. 59), but in the entire habitus. The Early Eocene

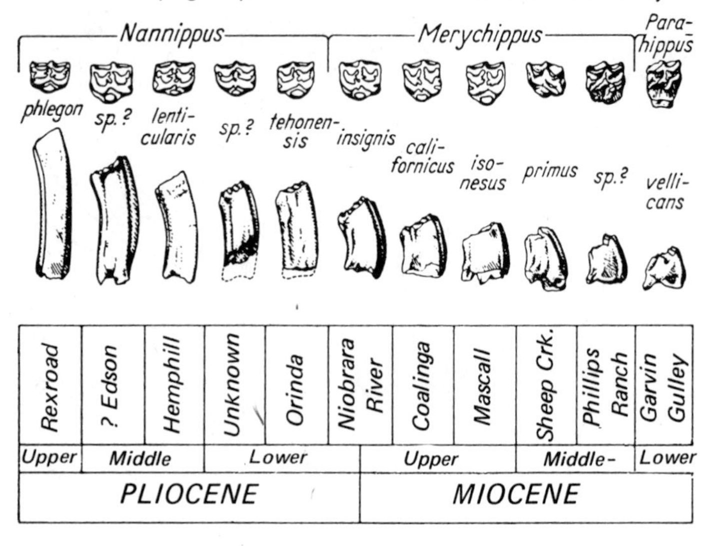

Fig. 59. Increase in crown height of molars (M sup.) in the Miocene-Pliocene horses of North America. Note constant increase. (After G. H. R. v. Koenigswald, 1968.)

equids were about the size of foxes, hoofed but with three or four toes, and, like the present-day Duikers of Africa, they were jungle-dwellers, at home in a tropical climate; they did not look in the least like a horse. They ultimately developed into the Miocene anchitherians which, along with *Hypohippus* and *Megahippus,* died out during the Pliocene epoch. These were three-toed animals about the size of a horse, and they were browsers. The line of development which led to the present-day one-toed horses separated from the main branch as long ago as the end of the Early Tertiary period, and, in contrast to the forest-dwelling anchitherians, showed a rapid increase in the

[3] In a single sequence of ancestors, the various members recovered from beds of different ages descend from one another. In a sequence of changes, this is not the case. Finally, we must mention the anatomical or morphological sequences, which are compiled from contemporary, but different, highly specialized species and are really intended simply to give some idea of the possible course of phylogeny.

crown height of the molars, associated with the change to feeding on steppe grasses, which began in the Late Tertiary period. A split occurred among the grazing equids, producing at least six different lineages, only one of which (genus *Equus* in the broad sense) has survived until the present. These are the true solid-hoofed animals. The hipparions, which died out in the Pleistocene age, were also grazers but not truly one-toed — they had three toes. The interplay between environment and adaptation, as the development of the horse shows, should not be understood to mean that the environment stamps the organism and hence exerts a direct influence upon evolution. Rather, changes in inherited characters induced by gene mutations lead, via the mechanism of selection, to improved adaptation to the changing conditions of the environment.

There are a number of other questions linked with the example of the horse, and we shall refer to them briefly. Thus, the rate of evolution is certainly not constant. It is just as likely to change within a lineage or to proceed at various speeds in different organs, as is clear from the development of the brain and jaw. The skull and jaw of the Eocene *Hyracotherium (="Eohippus")* enable it to be placed without any doubt among the equids; yet its brain (cast of the cavum cranii) is much more primitive, resembling that of early insectivores or marsupials. Comparative studies on the rate of evolution have shown that it was in general considerably faster in the mammals of the Paleocene, the earliest age in the Tertiary period, and in the Pleistocene age (Ice Age) than in the remainder of the Tertiary period (Eocene to Pliocene). This finding is attributed to the amount of empty living space then available. At the beginning of the Tertiary period there was a great deal of terrestrial living space unoccupied, owing to the extinction of the Mesozoic dinosaurs; a similar situation prevailed during the Ice Age because of the repeated alternation of cold and warm periods. The "horse sequence" is also an example of increase in size in the course of evolution (Deperet's law), since the geologically youngest species are always the largest and the geologically oldest the smallest. But the "horse sequence" is also an example of increasing specialization during phylogeny. This is equally true of the molars and the dimensions of the limbs. Naturally, such an increase in level of specialization or degree of adaptation restricts the evolution potential to some extent, and this is yet another factor in the controversy surrounding "orthogenesis." Usually, however, increasing specialization leads to extinction, as expressed in the law formulated by the American paleontologist E. D. Cope.[4] This law states that the least specialized species are those best able to survive. Increasing specialization is not only one of the most universal features of evolutionary events, it is also one of the conditions which most frequently precede extinction. Once again, we come back to the controversy surrounding *extinction.* By this we mean, of course, natural extinction, not extermination by the hand of man, although the latter is almost always merely an acceleration of the process of natural extinction which was not even possible before the introduction of firearms (e. g., extermination of Steller's sea cow or the South African quagga, a kind of zebra).[5].

Natural extinction is a biological process leading to the disappearance of a species or of an entire lineage. The disappearance of a complex of forms by evolutionary transformation into a new species, such as occurred in the ancestral sequence described above, cannot, of course, be called extinction. The reasons why species become extinct are no clearer today than when the phenomenon was first recognized, but it is clear that a number of conditions or factors tend to hasten the process. Changes in the macroclimate, predators, and disease have been suggested as causes, and attempts have been made to produce evidence in support of them. It is true that macroclimatic changes assist in eliminating a species, but the true reasons seem to have their roots within the organism itself. The most important precondition is over-specialization: thus, at the end of the Pleistocene, only the most highly specialized species became extinct — for instance, mammoth, woolly rhinoceros, steppe bison, giant deer, and cave bear — because they were unable to adapt to the new territory as it was altered by afforestation.

In connection with this discussion of the causes of extinction, we should also mention the occurrence of giant and dwarf forms. In many animal groups the emergence of dwarf forms seems to have preceded extinction. Examples are the dwarf elephants of the Ice Age,

---

[4] This "law of unspecialized species" was, however, formulated for the first time in 1855 by the British zoologist R. Wallace.

[5] In prehistoric times, with the exception of the flightless moas of New Zealand, which were exterminated by the Maoris, and of the giant prosimians of Madagascar, which were possibly exterminated by the primitive inhabitants of the island, no single game animal was exterminated by contemporary man, although he hunted numerous kinds of game.

belonging to the group of *Palaeoloxodon antiquus* of the isla
the Mediterranean; the Pleistocene dwarf elephants of Celebe
Late Pleistocene Siberian mammoth *Mammuthus primigenius ber
kius (=sibiricus)* which was smaller than the mammoth of the h
of the Ice Age; and the Pleistocene dwarf hippopotamus o
Mediterranean area and Madagascar. None of these survives to
Moreover, they inhabited regions where elephants and hippos ar
longer found. Is it possible to say that the appearance of dwarf fo
invariably precedes extinction? The answer to this question is
emphatic "no." In some extinct lineages, development came to an
in giant forms—e.g., titanotheres, giant sloths, giant armadil
giant ostriches, many dinosaurs, and giant scorpions—while in ot
lineages the forms were "normal" in the sense that terms like "giar
and "dwarf" are in any case relative.

In the course of phylogeny, however, not only greater specializ
tion but also more complex development occurs. Viewed from a
evolutionary angle, this complexity is not to be regarded as harmfu
since it leads to increasing emancipation from the environment. Suc
factors are warm-bloodedness, larger brain, reproductive process, etc
It was differentiation of this sort which eventually led to the emergence of man from among the primates.

There is yet another law of evolution—the "law of irreversibility," which is named after the Belgian paleontologist L. Dollo. This law, which states that development cannot be reversed, is no longer regarded as a law, but merely as a principle. According to O. Abel it should be taken to mean that once an organ has regressed completely in the course of phylogeny, it is unable to return to its earlier form. For example, a one-toed animal cannot go back to being a multitoed form. Changes during growth may, however, sometimes pass through ancestral development stages, and this can occasionally effect a kind of reversal of evolution, as J. Wiedmann recently observed in the "divergent forms" of ammonites, otherwise called heteromorphous ammonoids.

Examples of Phylogeny Attested by Fossils: Proboscideans, Bears, Lung Fishes, Ammonites

The fossil record has made the history of the proboscideans (elephants, mammoths, etc.) as familiar as that of the horse. There are at present only two extant forms of proboscideans—the Indian ele-

phant and the African elephant. They are found only in South Asia and Africa, although at one time they were was distributed all over the world and exhibited a rich variety of forms. The geologically oldest proboscideans are known from Eocene deposits in Africa, their primordial habitat. These were mammals about the size of a tapir, and they had neither a trunk nor tusks, merely a few elongated incisors, which evolved into tusks in later forms. The Oligocene *Palaeomastodon* had a slightly extended upper lip which, by the time of the Late Tertiary mastodons, had become a true trunk (Fig. 60). As the trunk grew longer, the skull also changed its shape to make room for the muscles of the trunk in addition to those of the jaw. The molars also experienced gradual changes, culminating in the complex structure of elephant molars which consist of numerous lamellae. This process reflected a change in feeding habits as compared with the proboscideans of the Tertiary period and resembled the events seen in equids, although in the latter this change took place in the Late Tertiary period. Contrary to earlier ideas, elephants evolved not from the stegodons, but directly from the Tertiary mastodons. Recent elephants must accordingly be regarded as the last survivors of a lineage of mammals which were once widely distributed over the whole of Eurasia, Africa, and North and South America.

No less informative is the fossil history of the bears (ursids). One of the geologically most ancient bears is the South American spectacled bear *Tremarctos ornatus*, which is not closely related to the other bears. The Malayan sun bear *Helarctos malayanus* and the sloth bear *Melursus ursinus* are also geologically ancient; the latter however, shows a number of secondarily acquired features of specialization. The brown bear *Ursus arctos* and polar bear *Ursus (Thalarctos) maritimus* are the youngest ursids, geologically speaking. The black bears, represented in Asia and North America by the collared bear *Ursus (Euarctos) thibetanus* and American black bear *Ursus (Euarctos) americanus*, have reached a less advanced evolutionary stage than the brown bears. The largest bears, the Late Pleistocene cave bears *Ursus spelaeus* of Europe and the Pleistocene shortfaced bears *Arctodus* (= *Arctotherium*) of America, are extinct.

Quite a different pattern of evolution is seen in the lung fish — the living representatives differ hardly at all from their Devonian ancestors (Fig. 61). There are three distinct lineages, all of which are restricted to the southern hemisphere: *Epiceratodus* Australia; *Proto-*

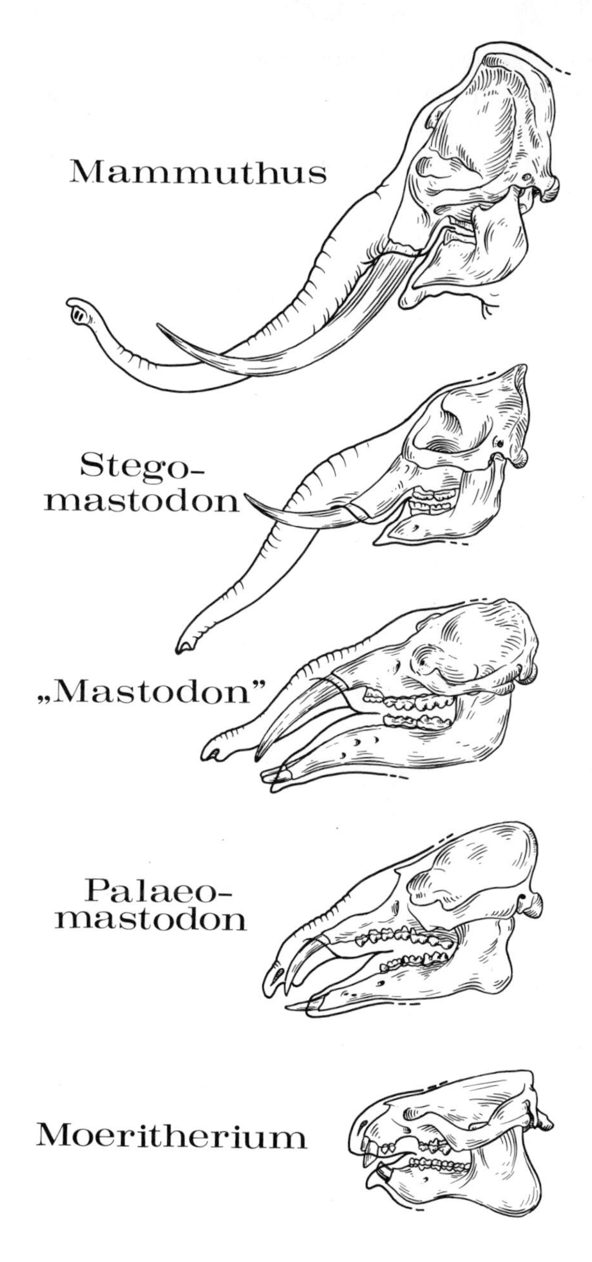

Fig. 60. Evolution of the proboscideans. How the trunk and tusks came into being. *Moeritherium*—Eocene; *Palaeomastodon*—Oligocene; "*Mastodon*" *(Gomphotherium)*—Miocene; *Stegomastodon*—Plio-Pleistocene; *Mammuthus*—Pleistocene. (After E. Thenius and H. Hofer, 1960.)

pterus Africa; *Lepidosiren* South America; originally, however, they were distributed throughout the world. The Australian lung fish hardly differs at all from Ceratodus of the Permian-Triassic period and may rightly be called a "living fossil" (see Chapter X). The

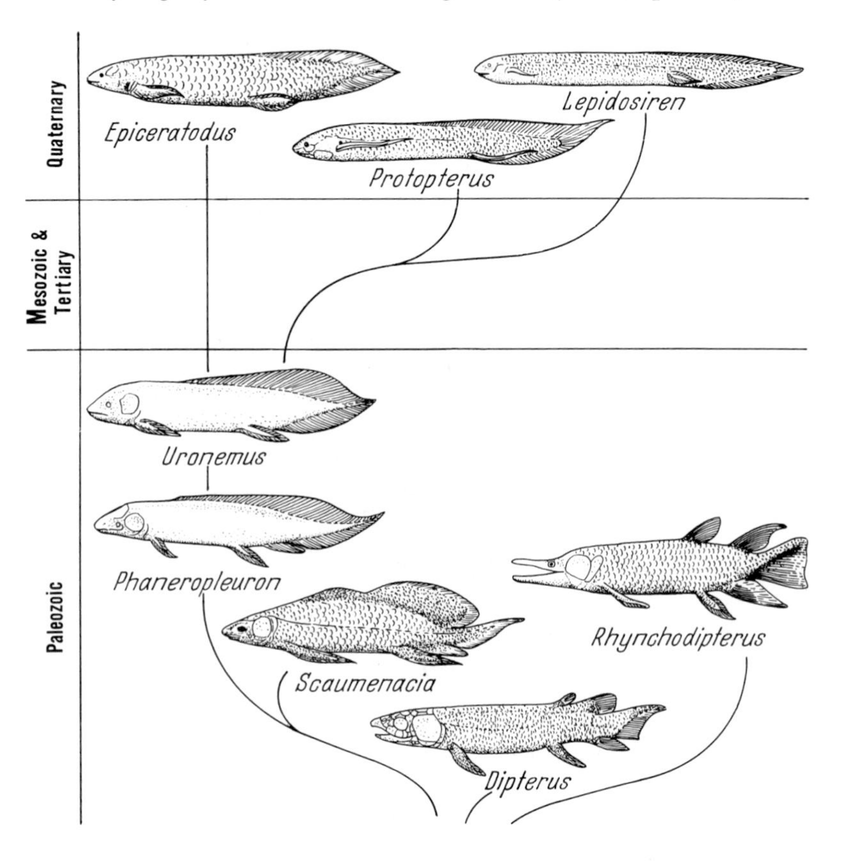

Fig. 61. Evolution of the lung fishes (Dipnoi). Great abundance of forms in the Paleozoic (Devonian). At the present time only the Ceratodontidae *(Epiceratodus,* Australia) and the Lepidosirenidae *(Protopterus,* Africa; *Lepidosiren,* South America) are extant.

Recent lung fishes of Africa and South America differ from their Paleozoic ancestors mainly in their much transformed pairs of fins, regression of scales, and their elongated, eel-type body. Closely related, they must have diverged as late as the Cretaceous period (see also Chapter IX).

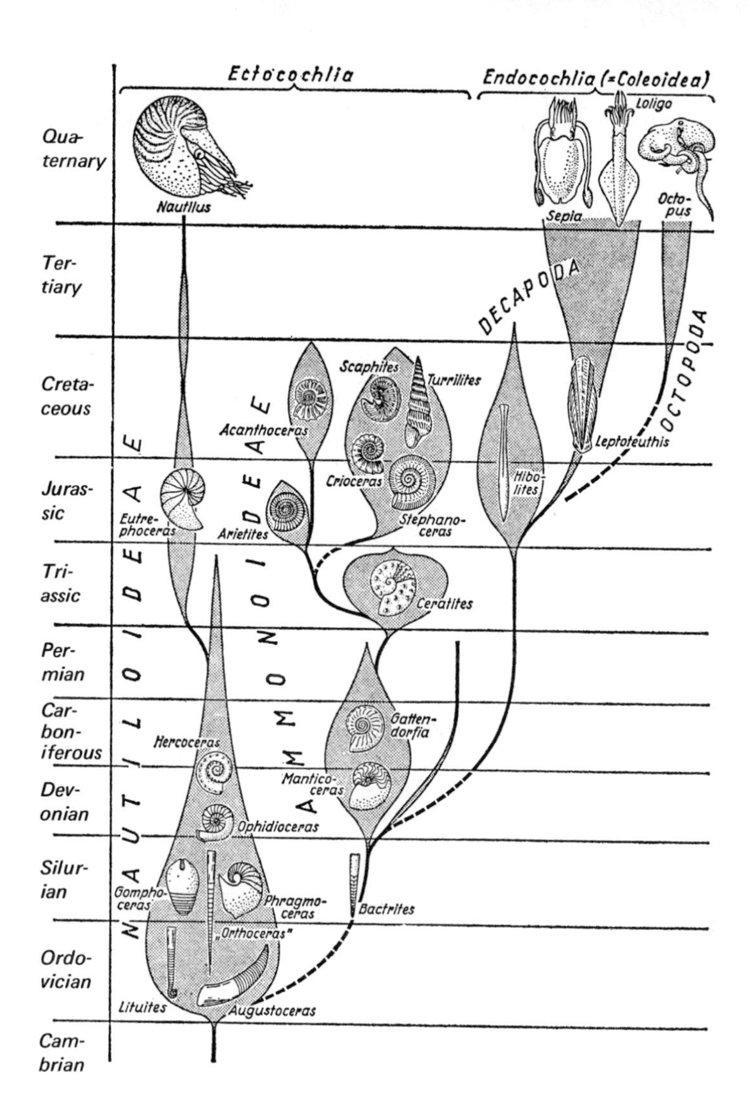

Fig. 62. Evolution of the cephalopods. Simplified scheme. Wide range of forms of Nautiloidea in the Paleozoic, the sole survivor being *Nautilus.* Note extinction of ammonites at the end of the Triassic and Cretaceous. Ammonites and "true" cuttlefish (Coleoidea) belong to the same group, the Bactritida. (After E. Thenius, 1965.)

Among the invertebrates we shall mention the ammonites, as their evolution is well known in its fundamental features, thanks to the abundant fossil material. Ammonites are shell-bearing cephalopods which, although they became extinct at the end of the Mesozoic epoch, resemble the present-day *Nautilus*. Ammonite shells, always planispirally coiled, are derived from the straight (orthoconic) forms of the Early Paleozoic era *(Bactrites)*, which lead still further back to the primitive Nautiloidea (Sphaerorthoceratinae). Worthy of note is the increasing complexity of the lobular lines marking the place where the septa are attached to the outer shell which, in the ammonites, progressed in the course of evolution from the simply vaulted goniatitic

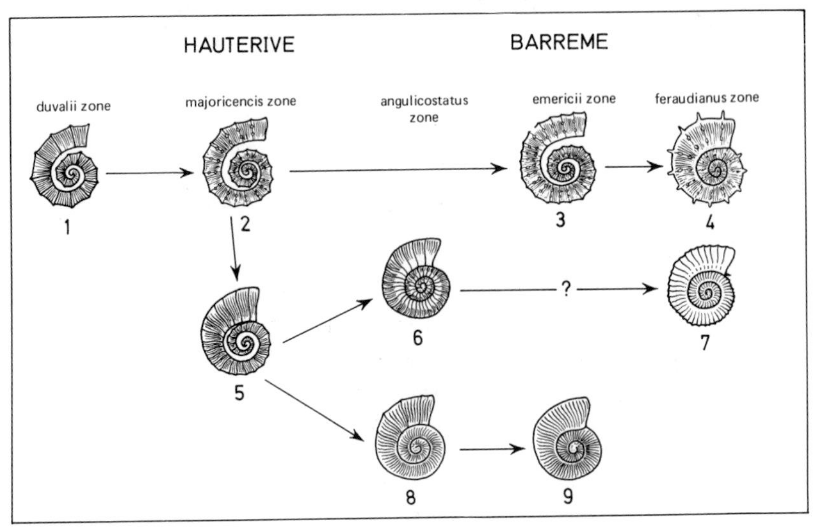

Fig. 63. Evolution of heteromorphic ammonites in the Lower Cretaceous. The Crioceratids show a tendency to return to coiling. Various subgenera of *Crioceratites:* 1—5 *Crioceratites,* 6, 8 and 9 *Pseudothurmannia,* 7 *Hemihoplites.* (After J. Wiedmann, 1969.)

(Devonian-Carboniferous) via the already unilaterally scalloped ceratitic (Permian-Triassic) to the complex ammonitic (Triassic-Cretaceous). Secondary simplifications do occur, however — the pseudoceratitic stage of the Cretaceous, for instance. The shell sculpture follows a similar course, the increasing complexity being apparent in the ribbing, simple and multiple knots, and so on, again with secondary simplifications. It is interesting that there is a break at the turn of the Triassic-Jurassic periods because all ammonites died out, with the exception of the phylloceratids and lytoceratids, the groups which dominate the Jurassic- Cretaceous ammonites (Fig. 62). Equally strange is the occurrence of divergent forms which have different types of shell—such as incomplete spiral rolling, irregular rolling, and partly unrolled shells—in the Later Triassic and Cretaceous periods. These divergent forms are limited to certain families, however (Fig. 63). There has been much discussion about why the ammonites (and other forms of life) vanished at the end of the Mesozoic era. But it should be noted that their disappearance was not sudden, nor did it occur simultaneously within the various groups. There are signs of a decline in the number of species and forms in many Mesozoic lineages before extinction finally came. The reason is generally stated to be a worldwide deterioration in climate, but other causes of a different kind are also thought to have contributed.

"Connecting links": Psilopsids, Monoplacophorans, Ichthyostegalians, Seymouria, Archaeopteryx

The cases we have so far described have provided some insight into the process of evolution but, apart from the therapsids, have not included an example of a "connecting link," i. e., a form transitional between two groups which are today widely divergent from each other. We should like to close this chapter with a few examples of this kind.

Let us first take an example from the vegetable kingdom. Today the club mosses (Lycopodiales), horsetails (Equisetales or Articulata), and ferns (Filicales) are three clearly differentiated systematic members among the continental flora, which are called pteridophytes, or spore-bearing vascular plants, by the taxonomists. They have in common the division into roots, stem, and leaves, but they have characteristic differences, not only in the form of the leaves — the lycopsids and horsetails have small leaves (= microphyllines) and the ferns large leaves (= macrophyllines). In the horsetails the relatively small leaves are arranged in whorls at certain nodes up the stem, thus producing the very visible subdivisions in the body of the plant which give them their alternative name, Articulata. Other differences concern the structure of the fruiting organs, that is, the position and conformation of the sporangia in the three groups.

Now let us consider whether paleontology can tell us anything about the origin and evolutionary connections of these three groups on the basis of the fossil evidence. The geologically oldest remains of land plants are found in the Silurian period, but it is not until the Devonian period that we have more complete specimens which also provide information about anatomical structure. Nevertheless, the first attempted reconstruction of such a plant was published over 100 years ago. These plants cannot be called either club mosses or horsetails, or even ferns, for they consisted of a "rootstock" which crept over the ground; they had no proper roots and their aerial organs were entirely leafless, frequently branching dichotomously and bearing small sporangia at the tips of the ultimate branches. They are called psilophytes, or naked plants, and the Lower Devonian, the period when they were most varied and widely distributed, is also known as the psilophyte period. Our knowledge of these geologically oldest land plants has increased a great deal in the last several decades — now some forms of psilophytes are even known to have had leaves. Thanks to silicified forms found in the Devonian of Rhynie in Scotland, we can now see the finer details of their internal structure — vascular bundles, stomata, and the content of the sporangia. Psilophytes are true vascular plants, unlike the algae, fuci, and fungi which, incidentally, are not divided into roots, axils, and leaves and are consequently called thallophytes (thallos = shoot); thallophytes were the only plants on Earth until the first land plants emerged in the form of psilophytes. The most significant evolutionary steps involve the formation of vascular bundles for the transport of the water required for assimilation and transpiration, supporting tissues (not needed in aquatic plants) and the cuticle, or outer covering of the leaf, which prevents dehydration, fungal infection, and so on. Only these advances made it possible for plants to emerge from water on to the land. The psilophytes also form the morphological connecting link between thallophytes and pteridophytes, although the *Devonian* psilophytes cannot be regarded as the direct ancestors of the club mosses and horsetails, as even in the Devonian there were already some primitive forms beginning to diverge from the psilophytes. Although we cannot claim that the majority of psilophytes form a connecting link, the very fact of their existence proves that there must once have been such plants.

Now let us pass on to the animal kingdom. Dish-shaped shells from early Paleozoic deposits have been known for a long time. Because of their similarity to the modern conical-shelled limpets (patellids), these shells were simply thought to be the remains of limpets. It was the mollusc specialist W. Wenz, who first observed in 1938 that these shells were not, in fact, true limpets, but snails that were basically unlike any other gastropod. He consequently placed them in an order of their own: Tryblidiacea. In 1940 the Swedish snail specialist Ohdner elevated them into an independent class, Monoplacophora, thus putting them on the same systematic level as snails, lamellibranchs, etc. On the inside of the shells of Tryblidiaceae there are, instead of one horseshoe-shaped depression where the muscle was attached, as is characteristic of limpets, a number of such impressions which Wenz thought might be due to segmentation, which seems probable for the worm-like predecessors of molluscs. This assumption was gratifyingly borne out only a few years ago by the fortunate discovery of living monoplacophores (*Neopilina galatheae* and related forms, see p. 170). The segmentation of these Recent monoplacophorans extends not only to the connective musculature and gills but right down to the excretory organs (nephridia) and nervous system. However, this segmentation is not actually a true or coelom-metamerism but merely a regulatory one, since all the organs mentioned are not present in equal numbers. Although their Precambrian ancestors are unknown, the Early Paleozoic monoplacophorans nevertheless provide confirmation that the original hypothesis of molluscs deriving from jointed worms is not true.

The situation is much more favorable for vertebrates, since their entire evolution occurred in the Postcambrian epoch and there is consequently plenty of fossil evidence. In this case, too, recent discoveries and studies have made a decisive contribution to our knowledge.

The crucial advance was without question the discovery of the geologically oldest amphibia, which have various characteristics previously known only from fishes and can thus to a certain extent also be looked upon as "connecting links." These amphibia were collected from the Old Red Sandstone (= Upper Devonian) of East Greenland in 1929 by Swedish paleontologists. In 1931 the young Säve Söderbergh published a report on the skull remains, calling it *Ichthyostega*, which means, more or less, "fish roof." Was it a fish, or was it a quadruped already? Subsequent expeditions found further material which proved that *Ichthyostega* was indeed a quadruped. But it was not until quite recently that E. Jarvik was able to show that *Ichthyostega* in its adult state possessed a true fishtail and vestiges of gill covers and of the second dorsal fin, as well as a system of lateral-line canals. Thus, *Ichthyostega* combines the characters of fishes and amphibians (Fig. 64). The structure of the skull and shoulder girdle, too, brought further illumination. But many characters of *Ichthyostega* (e. g., position of external nostrils) are already too specialized for it to be considered a direct ancestor of the other amphibians, although this does not destroy its fundamental evolutionary importance, particularly when we consider the mosaic mode of evolution mentioned on p. 96. The ichthyo-

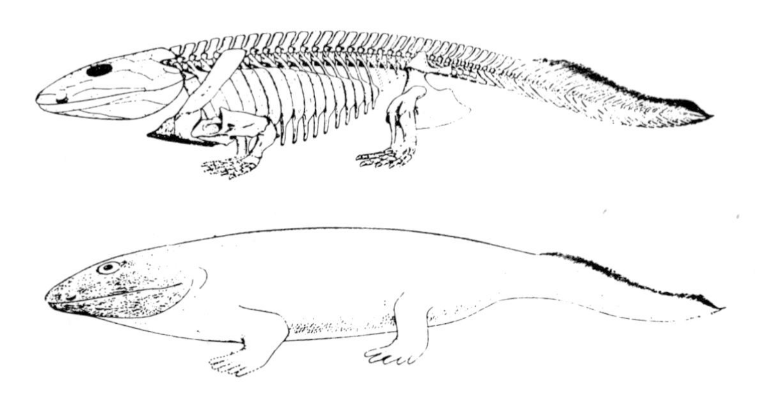

Fig. 64. The oldest land vertebrate *Ichthyostega* from the Upper Devonian of Greenland combines features of fishes, tail and dorsal fin, remains of gill cover, and so on, with those of amphibians. Reconstruction of skeleton and habitus. Length about 1 m. (After E. Jarvik, 1960.)

stegalians show how amphibians might have developed from the crossopterygians (Rhipidistia), and they are not only the oldest four-footed animals but also the most primitive land vertebrates.

The same sort of intermediate position, this time between amphibians and reptiles, is occupied by the genus Seymouria found in the Texas Permian; this genus combines in its skeleton the characters of both, so that some systematists place it among the Amphibia and some among the Reptilia. However, the paleontological record does not permit the differences between the members of these two groups of vertebrates to be clearly distinguished, since the decisive features are thought to have been the form of the skin (without covering, slimy or horny) and the manner of reproduction (larval stage in water, or eggs in shells laid on land). But in spite of this, paleontology can make a definite contribution to solving this problem. Once it was supposed that the first stage of the transition from aquatic to terrestrial living occurred with the emergence of reptiles, and that the amniote egg[6] was "invented" only after the emergence from the water was complete; however, on the basis of fossils from the Texas Permian, A. S. Romer made the well-founded assumption that the amniote egg came into being first and that only after that did the animals emerge from the water. For the reptiles found associated with the geologically oldest amniote egg, the cotylosaurians and the pelycosaurians, are still completely adapted to life in the water — they have aquatic skins, are fish eaters, and so on. The development of the amniote egg was the precondition for the final conquest of dry land by the vertebrates.

The relationships are more clearly seen in the connecting link between reptiles and birds. Until 1861 no remains of birds of the Mesozoic era had been found, and then within the span of 20 years came the sensational discoveries of primeval birds from the Solnhofen platey limestones of the Upper Jurassic period. The first skeleton to be found, which did not include the skull (London specimen), was preceded by the impression of a feather; the second skeleton (Berlin specimen) had the skull as well (Fig. 65). Controversy raged among scholars for a long time as to whether this was *still* a reptile or *already* a bird, or simply an intermediate form which could be regarded as an ancestral bird. These differences of opinion arose from the combination of morphological features possessed by these primeval birds. An analysis of the remains of these two specimens of *Archaeopteryx lithographica*[7] and of a third which had been recovered in the meantime indicated that they combine primitive and typically reptilian characters (toothed jaws, os sacrum consisting of six vertebrae, long tail spine, development of the brain, and free clawed fingers) with genuine bird features, such as feathers, furcula, os pubis pointing to the rear, and

[6] The amniote egg, which possesses some of the characteristics of reptiles, birds, and mammals, has not only a resistant envelope (serosa and amnion) but also an embryonic urinary sack (allantois) which enables the development of the embryo to take place out of the water. Although the serosa, amnion, and allantois have not been preserved in fossil eggs, it may be concluded from the presence of the resistant external shell that these must have been there originally because without them no development could have taken place out of water.

[7] The classification of the various remains into species and genera which was made at that time was, as later studies by G. de Beer and F. Heller have demonstrated, based essentially on individual differences due to age—earlier or later closing of the various sutures, etc. In fact, the remains all belong to a single species.

Fig. 65. Skeleton of "primordial bird" *Archaeopteryx lithographica* H. v. M. from the Upper Jurassic platey limestone of Solnhofen. Berlin specimen, about one-seventh natural size.

opposed big toes. Reptiles or birds? To argue about such matters is utterly pointless, as has already been shown in connection with the therapsids.

The derivation of birds from reptiles (pseudosuchians) is no longer seriously doubted. Drawing a line between them is simply a matter of convention. If we take the possession of feathers as the prime characteristic of birds, then *Archaeopteryx lithographica* is indeed a bird. It is a primitive sort of bird, but a bird nevertheless, not to be regarded as simply a convergence phenomenon among the reptiles. There are many other features which confirm this opinion, but which cannot be discussed within the scope of this book — the fact that the eye has fourteen scleral plates and the diastaxy of the wings in *A. lithographica;* also the "*Archaeopteryx*" tail of bird embryos, etc. At the same time the taxonomic significance of some of the characters becomes clear. If *Archaeopteryx* were the only species of birds known to us, they would presumably be classified as aberrant reptiles and not as the representatives of a whole new class of vertebrates, namely birds.

The growing body of knowledge about such evolutionary transitional forms has, however, shown that various categories traditionally classified as units are of "polyphyletic" origin, or rather of polygenetic or polyradical descent. They have developed independently of one another from various radical forms, which means that vertebrate classes hitherto regarded as natural units are not entirely so and ought, if the system is also to be a true expression of our knowledge of phylogeny, to be broken down further. This applies not only to the amphibians (urodeles and frogs), but equally to the reptiles.

Connections between the Evolutionary Development of Plants and Animals

If we take a general view of evolution in its broad outlines, we can observe a remarkable, although theoretically predictable, obedience to some general principles. Plant development precedes that of animals and is to some extent the pathfinder. This becomes particularly clear if we compare the limits of the Paleophytic and Mesophytic, or those of the Mesophytic and Cenophytic eras with those of the Paleo- and Mesozoic and Meso- and Cenozoic eras. Their boundaries do not coincide at all, the limits based on plants being always half a period earlier; between Lower and Upper Permian as against Permian and Triassic, or between Lower and Upper Cretaceous as against Cretaceous and Tertiary. The plants were the first to invade dry land and, by their activity of photosynthesis, they not only prepared a source of food for herbivores, they also indirectly opened the road to carnivores. The development of flowering plants (angiosperms), steppe grasses, and so on produced a subsequent "echo" in the insect world — pollination by insects, one-toed ungulates, etc.

Origin of Life: Precambrian Fossils

To conclude this chapter, let us say something about the problem of the beginning of life on Earth and hence about the geologically oldest fossils. Unfortunately, paleontology is for several reasons unable to solve the problem which is of prime importance to all life scientists—the problem of how life began on Earth. Nevertheless, the oldest fossils do provide some interesting pointers to their organization and appearance in time, as well as to the evolution of early forms of life.

The search for the geologically oldest traces of life on Earth is as old as the study of the oldest rocks. In the course of the last several decades a number of discoveries from Precambrian rocks have been described. A critical examination of these "Precambrian fossils" has shown, however, that for many remains neither their organic nature nor their Precambrian age is all that certain. Thus, despite all this new material, there is a striking contrast between the dearth of animal fossils in the Precambrian period and their abundance in the Cambrian period. This section of the fauna is currently the subject of lively discussion. According to O. H. Schindewolf, it is a genuine biological phenomenon, which is determined by something more than the condition and degree of metamorphism of the rocks in question. If the organisms had existed then, the rocks would at least bear imprints of them. Attempts to explain the poverty of fossils in the Precambrian rocks by their predominantly continental facies, or by relating them to a transition from planktonic to benthic organisms about the end of the Precambrian and beginning of the Cambrian period, have come to naught. More credible is M. F. Glaessner's hypothesis which postulates that living organisms only slowly acquired the capacity to precipitate calcium, and hence the necessary condition for fossilization.

This hypothesis is supported by various groups of organisms whose skeletons consisted initially of organic substances, such as horn or chitin, but were gradually replaced by calcified tissues — e. g., Foraminifera, brachiopods, and possibly corals.

In recent years systematic searches for fossils have been made in the Precambrian rocks from the old continental masses, such as the Canadian, Brazilian and Baltic shields, South Africa, and Australia. Particular attention has been focused on cherts in which microfossils, too, are very well preserved.

Fig. 66. Fossils from the Precambrian of Ediacara, South Australia. 1—4 *Spriggina floundersi* Glaessner; 5—6 *Parvancorina minchami* Glaessner; 7—8 *Tribrachidium heraldicum* Glaessner and Daily; 9 Problematicum (Siphonophore?). All reduced. (After M. F. Glaessner and B. Daily, 1959.)

The geologically oldest fossils are found in rocks more than 3 billion years old — e. g., the Fig Tree formation in Swaziland. These fossils are without exception the remains of unicellular and prenucleate organisms, principally bacteria *(Eobacterium)* and blue-green algae (Cyanophyceae). It is questionable whether these organisms were capable of photosynthesis, although chlorophyll derivatives, at any rate, have been found in rocks 2 billion years old. The oldest nuclear (eukaryotic) organisms, green algae *(Glenobotrydion)*, are found in the Late Precambrian rocks of Australia (Bitter Springs formation).

A fauna from the sandstones and cherts of the Latest Precambrian period of South Australia (Ediacara formation) has been described in the last few years. Composed mainly of organisms without hard parts, it can only tentatively be given a systematic classification. According to Glaessner, these fossils, found exclusively as impressions, are primarily jellyfish-like and worm-type organisms: *Spriggina*, *Dickinsonia*, sea pens — *Rangea* and *Pteridinum*, presumably from mud oozes, and invertebrates of a wholly unfamiliar type: *Parvancorina* and *Tribrachidium* (Fig. 66). Similar forms have been described from Southwest Africa and Britain, but quite recently the remains ascribed by Glaessner to sea pens have been identified as tunicates. At the very least, these discoveries show that our knowledge is still slight when it comes to Pre- and Eocambrian organisms and that the roots of the various phyla of animals are to be sought in the Precambrian.

VII. Trace Fossils

In 1935, O. Abel published a comprehensive and richly illustrated work entitled *Vorzeitliche Lebensspuren* (Trace Fossils) which, in the author's own words, set out to be neither a handbook nor a textbook on the subject. It contains a wealth of more or less fully discussed examples. Since that time much new material has been collected, both by "contemporary" paleontological studies (see p. 66) mostly carried out by the research institute at Senckenberg am Meer, Wilhelmshaven, created by R. Richter, and by intensive analysis of trace fossils. This work has further emphasized the importance of trace fossils; a whole chapter is therefore devoted to them here.

What are Trace Fossils?

What does the paleontologist mean by "trace fossils"? As already briefly indicated on p. 13, trace fossils comprise all possible traces which throw any light upon the activity or way of life of the creatures that made them. Mostly they are simply, as the name implies, traces which are available for analysis by the paleontologist. Direct observations of the manner of life of fossil organisms or experiments with them are denied to the paleontologist for ever. But the following examples will suffice to show how far it is possible to go in evaluating the way of life of fossil organisms on the basis of these life traces.

Inorganic traces, in the form of flow marks, ripples, raindrop impressions, etc., are called marks in order to distinguish them from organic traces, or trace fossils. The distinction between traces and marks has proved useful in paleontology and geology, since certain trace fossils possess great practical importance for the geologist, as will be explained. Trace fossils comprise movement traces—tracks, creeping and burrowing traces; feeding traces—food remains, gastroliths and coprolites; dwellings — burrows and earthworks, holes of boring molluscs; life communities—symbioses, parasitism, etc.; traces of reproduction — egg cocoons, nests; and pathological phenomena — broken bones, spondylarthroses, rachitis, caries, etc.

We shall select just a few of the many kinds of trace fossils in order to show how they may be interpreted (for example, tracks) and to indicate their importance for various related disciplines: salinity for geology; bathymetry of marine sediments for paleogeography; and hyena feeding traces for prehistory.

Analysis of Tracks

First, the analysis of tracks. In most cases the originator of the fossil tracks is unknown; it is very seldom that the creature meets its end at the end of the track and becomes fossilized there, like the horse shoe "crab" *Mesolimulus* in Fig. 67. The creeping tracks of such "crabs" were originally attributed, because of their resemblance to those of land vertebrates, to amphibians, flying reptiles, and primitive birds. Some of the best known tracks are those described under the name *Paramphibius* from the clay shales of the Upper Devonian period in Pennsylvania, which were made by "crabs" related to *Protolimulus.* The traces known as *Kouphichnium lithographica* and *Protornis bavarica* from the Solnhofen platey limestones are the creeping traces of *Mesolimulus.*

In the same way as the huntsman studies game tracks and draws conclusions from them about the originator and its behavior, so the

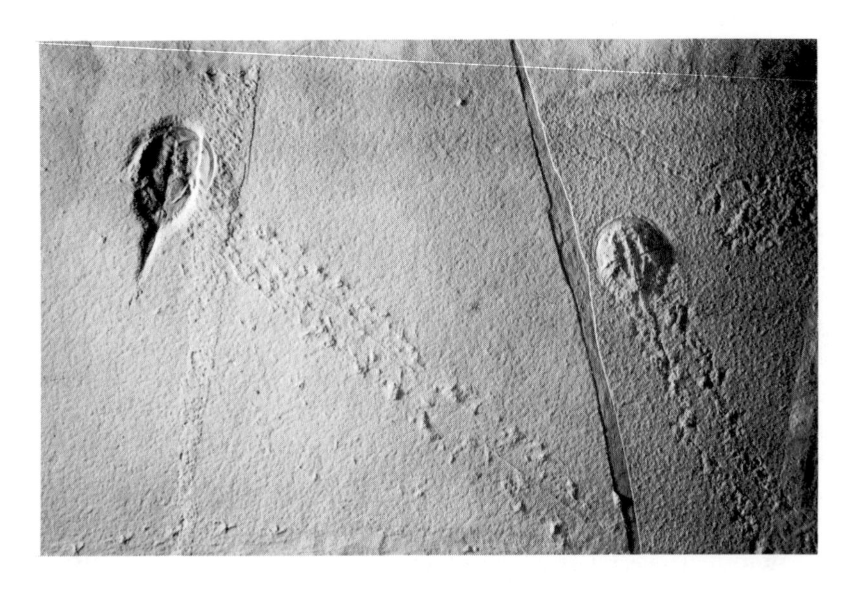

Fig. 67. Trace fossils and the organisms that produced them: king "crab" *Mesolimulus walchi* (Desm.) which died and became buried at the end of the track. From the Upper Jurassic platey limestone of Solnhofen (Bavaria). Original in Senckenberg-Museum, Frankfurt (Main).

paleontologist can analyze his fossil track and draw conclusions about the organism that produced it and what it was doing at the time. In this sense one can speak of *paleoethology,* since in appropriate cases some conclusions about the behavior of the primitive organisms can be reached.[1] This is not to say that behavior can be assessed from tracks alone. Trace fossils of any kind are used for such interpretations. However, let us return to the analysis of fossil tracks.

The tracks of chirotherians are found in large numbers in the Chirotherian Sandstone of the earlier continental Triassic period of

[1] In the paleontological literature, the expression paleotaxiology has come to be accepted for analyses of this type. Richter even speaks of paleopsychology.

Germany. For more than 100 years descriptions have also been available from places other than Germany, but it was not until W. Soergel's careful analysis that a proper assessment of their "makers" from a taxonomic viewpoint was attained. The chirotherian tracks are present mainly as reliefs on the underside of the beds of sandstone (Fig. 68). This phenomenon also makes it clear how such traces may be pre-

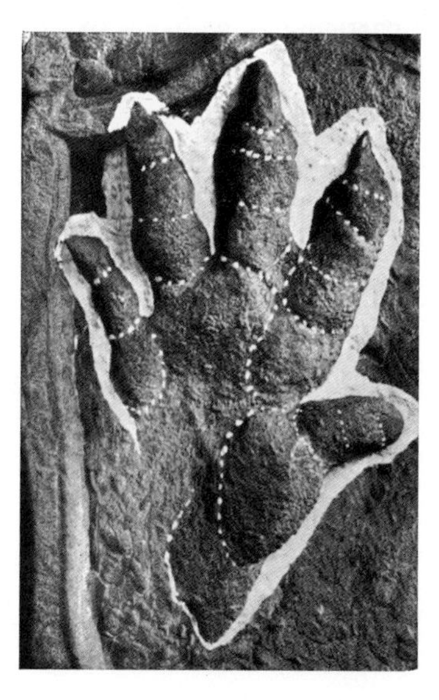

Fig. 68. Track impression of left hind foot of *Chirotherium barthi* from the Early Triassic of Hildburghausen. Relief highlighted by outlining in white and by dotted lines added between toes and central foot pads. Fifth toe resembles big toe. Magnification about one-third. (After W. Soergel, 1925.)

served: the tracks themselves were made in soft, clayey soil which, as the desiccation cracks show, dried out afterwards; the tracks were then covered by a layer of fine sand. This change in the type of sediment brought about the preservation of the tracks and also explains why they are nearly always in the form of a relief. Soergel was able to reconstruct from the tracks alone (there were no skeletal remains of their "makers") not only *Chirotherium* itself in the form of a block model (Fig. 69 a), but also to a considerable extent its position within the order Reptilia — he placed it among the pseudosuchians. His results were later brilliantly confirmed by discoveries made by F. von Huene in deposits of the same age in South America of the genus *Prestosuchus*, and more recently in the Southern Alps of *Ticinosuchus* by B. Krebs (Fig. 69 c). It should be added by way of explanation that the so-called phalange formula (number of bones in each digit), which plays an important part in reptilian systematics, can, of course, be deduced from the footprints (Fig. 69 b). Further, the tracks themselves and the pattern they formed (i. e., no marks suggesting dragging of the belly or tail) and in some particularly well-preserved imprints the type of skin, were also helpful in reaching a conclusion.

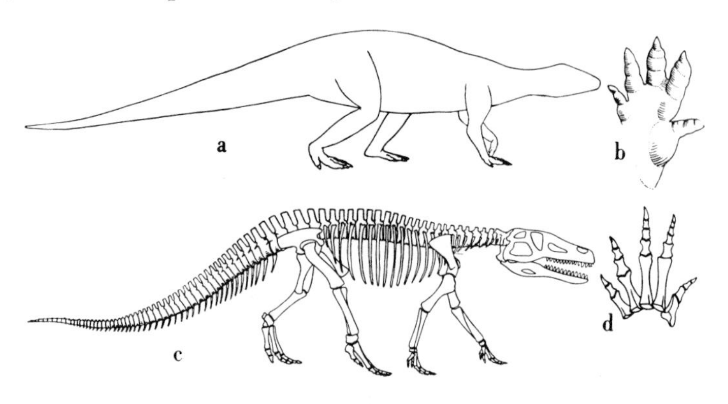

Fig. 69. a. Block model of *Chirotherium* reconstructed from tracks by W. Soergel; b. track impression of hind foot; c. skeleton of *Prestosuchus chiniquensis* v. Huene from the Triassic of South America, a reptile related to *Chirotherium;* d. right hind foot of *Prestosuchus.* (a—b after W. Soergel, 1925; c—d after F. v. Huene, 1942.)

The soil type may have played an important role in the formation of tracks by the chirotherians, and indeed, by vertebrates in general, but this fact is even more crucial when it comes to the tracks of invertebrates. Such tracks can look entirely different according to the amount of moisture in the ground. The movement traces of invertebrates include creeping traces, which sometimes lead to real burrowing traces and sometimes to feeding traces, although it is often difficult to distinguish between these. The traces of marine invertebrates are also important to the geologist because they make it possible to distinguish the top of a layer from the bottom; this distinction can be extremely important in rock series which have been subjected to strong tectonic disturbances.

One rock series of this kind is the flysch, a characteristic sequence of sandstones and limey shales of the Cretaceous and Early Tertiary periods, found in Switzerland, Austria, Czechoslovakia, Poland, Italy, Yugoslavia, and other countries. The flysch has long been famous for its rich abundance of trace fossils (Fig. 70), although it does not contain so many macrofossils. Very thorough analyses carried out by A. Seilacher with the assistance of "contemporary" paleontological

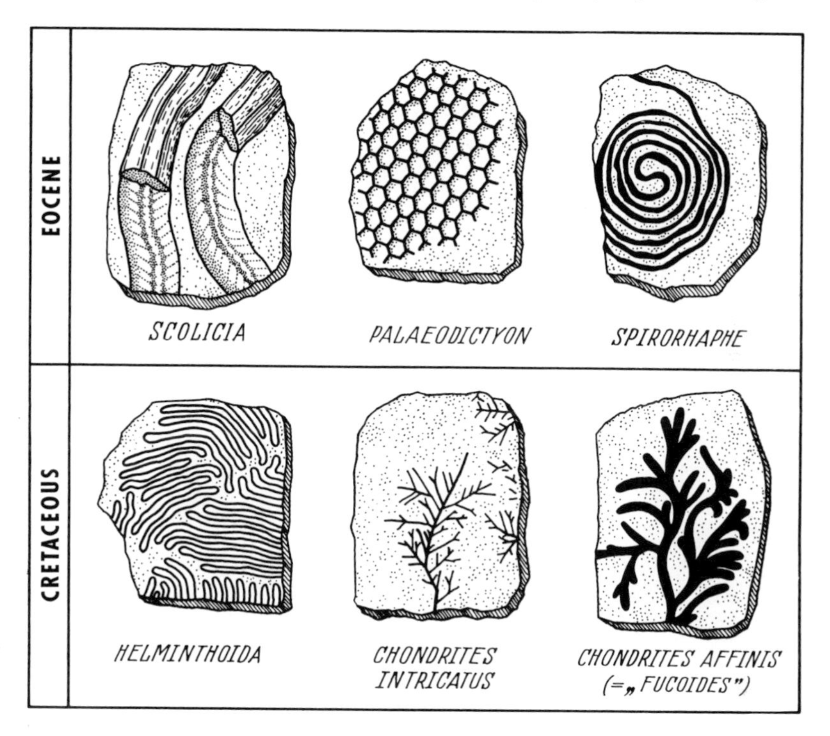

Fig. 70. Characteristic trace fossils of gastropods *Scolicia* = "*Palaeobullia*" and worms *Helminthoida, Chondrites, Spirorhaphe* and *Palaeodictyon* from the Upper Cretaceous and Eocene flysch of Europe. Dependence of trace fossils upon sediments as a cause of stratigraphic differences. Reduced to scale.

studies have shown that the deposition of the flysch, contrary to earlier opinion, did not occur in tidal areas, but at considerable ocean depths; thus deposition did not occur in a shallow sea. This finding is confirmed by lithological analysis (e. g., graded bedding resulting from turbidity currents) and by the microfossils. Here we can only briefly indicate the important implications this discovery held for paleogeographers and students of tectonics.

Within the scope of this volume we cannot go into great detail about trace fossils in the flysch, but we should just mention that the most frequent traces (at least, in the Mesozoic portions) — namely, the fucoids and chondrites — were originally assumed to be plant remains because they look like plants. But they are, in fact, the burrows of worms which lived in the mud, as is clearly shown by the form and position of these traces. The same is true of the equally highly characteristic helminthoids, known in the paleontological literature as "guided meanders" because of their characteristic shape. These are the feeding burrows of worms which lived in the sediment, as similar worms do today. The course of such burrows is basically determined by thigmotaxis (the urge to come together) and phobotaxis (the fear of touching); the urge to touch, of which the first trail is the expression, determines the course of subsequent trails, because the fear of touching prevents the trails from crossing. The practical value of these "guided meanders," which always occur in particular horizons, lies in the fact that these "sediment-feeding" worms always made the best possible use of the layer in which they lived. As Seilacher has shown, from the Paleozoic to the Mesozoic eras it is possible to observe an increasing perfection in these traces, which enabled their makers to utilize each unit of a layer more intensively.

Other characteristic flysch trace fossils can be attributed to the activities of snails which lived at the bottom of the sea or in the sediment and accordingly left different sorts of traces: *Scolicia* = *Palaeobullia* and *Subphyllochorda*. The origin of many trace fossils is still under discussion today — for instance, the pattern described under the name *Palaeodictyon problematicum;* however, this pattern has recently been described as feeding traces.

Although comparison with Recent traces is of the greatest help in interpreting fossil traces, the Recent traces are often concealed from us because they are made by dwellers in the sediment on the deep sea floor. Moreover, Recent traces may be difficult to recognize in the sediment, and diagenesis may result in trace fossils being clear but different in appearance from their Recent equivalents, quite apart from the fact that many of the species involved in their creation have long been extinct. Such circumstances make correct interpretation difficult, if not impossible, so that some errors are bound to occur. Thus, at one time the impressions found in the Solnhofen platey limestones — made by the shells of ammonites which drifted in the water, touching the sea floor only briefly and sporadically — were taken for the swimming traces of crossopterygians and sea turtles; this error was pointed out by A. Seilacher.

Hyena Feeding Traces and their Interpretation

Now let us consider some types of feeding traces. We already mentioned (p. 18) food remains found in the stomach or mouth of the Pleistocene mammoths. Even more remarkable are the feeding traces of hyenas. Pleistocene deposits often contain large collections of bones and teeth — this is the case in the Devil's hole near Eggenburg (Lower

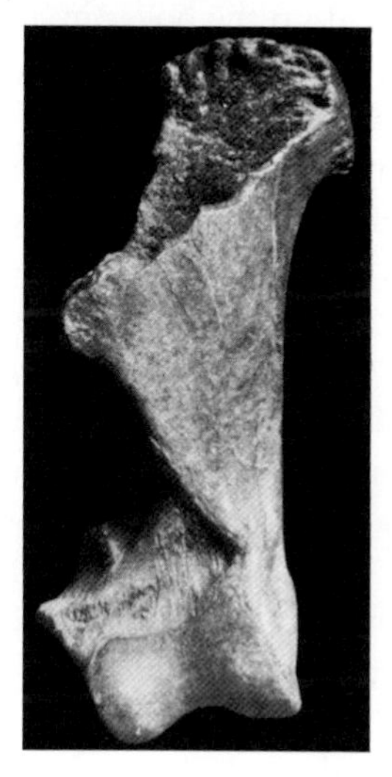

Fig. 71. Feeding remains of Pleistocene cave hyena *Crocuta spelaea* (Goldf.). Upper foreleg bone of woolly rhinoceros *Coelodonta antiquitatis* (Blum.), about one-sixth natural size. (After H. Zapfe, 1939.)

Austria), where excavations uncovered a rich hoard of Late Pleistocene vertebrate fauna. One of the striking things about this hoard was that the leg bones of the larger ungulates — woolly rhinoceros, wild horse, steppe bison, and giant deer — were almost all broken into pieces. The damage to the various bones, or their remnants, is so characteristic (Fig. 71) that we can exclude the possibility that it happened by chance. The conditions under which they were found indicated that the bones were already fragmented before they were buried. Similarly damaged remains of bones had been found in Pleistocene caves in England and Germany more than a century ago, and their fragmentation was usually attributed to human activity. The highly characteristic marrow-bone remains are referred to in the literature as "bell scrapers," "Kellermann buttons," and the like. However, quite early on, voices were raised in favor of interpreting these damaged marrow bones as the feeding remains of hyenas that were represented in Europe in the Late Glacial by the cave hyena *Crocuta spelaea.* Feeding experiments carried out by H. Zapfe of Vienna with hyenas in zoos showed beyond doubt that the Recent spotted hyena *Crocuta crocuta* — hence, a relative of the fossil cave hyenas — always crunched up the bones in a quite characteristic manner in order to get at the marrow. Instinctively hyenas always crunch off the less resistant portions at the end first, until eventually only the characteristic "bell scrapers" and "Kellermann buttons" remain. This contemporary paleontological research showed that it was not the activities of paleolithic man but those of cave hyenas which produced the bone fragments. That they are indeed the feeding traces of hyenas is clear not only from toothmarks on the bones themselves but also from the presence of the actual skeletal remains of these hyenas and their distinctive coprolites, or fossil excrement, which have been preserved as a result of their high calcium content. The remains are very typical of Pleistocene hyena lairs. This fact and the discovery of skeletal remains of newborn cave hyenas indicate that the hyenas probably had their home in these caves — at least, for some of the time — and dragged the bones of their victims, possibly even the complete legs, into their lairs in order to devour them there.

As is shown by the recovery of fragmented bones from the Pliocene clays of Pikermi in Attica (Greece), the Late Tertiary hyenas had already learned the trick of cracking open marrow bones with their teeth.

These investigations led to another fundamental conclusion, although those doing the research did not realize it because they were not trained in paleontology: the remains of the "osteodontoceratic culture" (so named by R. Dart) of the South African australopithecines are simply the feeding traces of hyenas. The australopithecines were, furthermore, not the antelope hunters whose remains are well represented in Makapansgat, Transvaal. Fig. 72 sets the feeding remains of Recent hyenas and the "tools" of the "osteodontoceratic culture" of the australopithecines side by side so that the reader may have the opportunity to form his own opinion. Let us mention in passing that excavations that have yielded australopithecines have also produced skeletal remains of hyenas and coprolites. A true evaluation of the

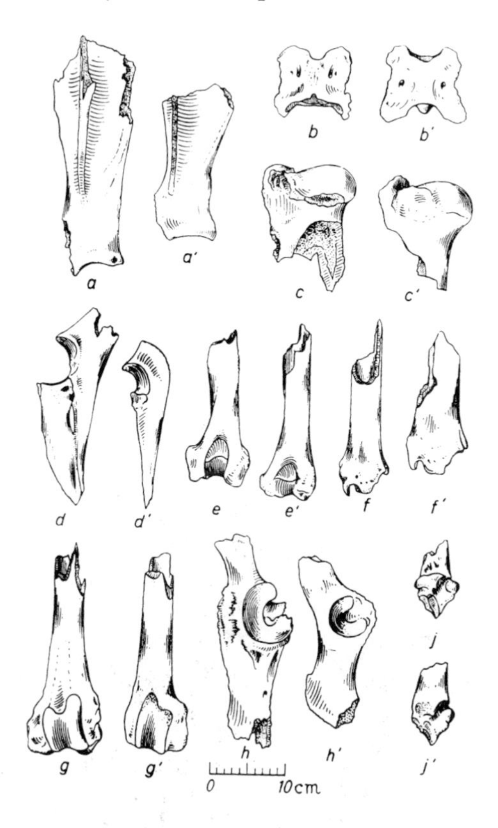

Fig. 72. Feeding remains (vertebrae and leg bones on ungulates) of Recent spotted hyena *Crocuta crocuta* (Erxl.) (a—j); presumptive tools of the "osteodontoceratic culture" of the australopithecines from Makapansgat, Transvaal (a′—j′). (After E. Thenius, 1961.)

"osteodontoceratic culture" is important because the australopithecines are the geologically oldest hominids, and hence the earliest type of humans. Furthermore, the ethnologists, in contrast to the anthropologists (see p. 70), regard as humans (Hominidae) only tool-makers, the tools being formed for a particular purpose which cannot be attributed to a purely random use. The makers of the stone artifacts found in South and East African locations are assumed to be more modern hominids, "*Telanthropus*" of Kromdraai and *Homo habilis* of Olduvai. However, this does not mean that the australopithecines did not use tools of wood or bone.

It should also be mentioned that many Early Stone-Age cave-bear hunting centers containing mass remains of the cave bear *Ursus spelaeus*, or the "protolithic bone cultures," have not stood up to unprejudiced analysis. The caves were the home of the cave bears, providing shelter for hibernation, birth, and death; the remains of bones and teeth, which had been supposed to be the tools of paleolithic man, accumulated without human assistance. The smoothing of the bone splinters, which had been regarded as a sign that they were used for some purpose, is, according to F. E. Koby, attributable to "dry scouring" (charriage à sec) caused by centuries of residence by cave bears. Moreover, the "Kiskevely blades" are merely splinters that broke off the canines of older bears in a perfectly natural manner because of their predominantly herbivorous diet.

Fig. 73. Feeding marks of crabs (pagurids or spiny lobsters) on the shells of fossil gastropods *Clavilithes parisiensis* (Mayer-Eymar) from the Eocene of the Paris Basin. Reduced. (After H. Zapfe, 1947.)

The next example concerns the feeding traces of invertebrate animals. The deposits of the Paris and Vienna Basins, which date from the Tertiary period, have yielded enormous quantities of marine invertebrates resembling in their variety that of the present-day inhabitants of the Indian Ocean or the Mediterranean Sea. Many of the fossil shells show characteristic damage to the mouth opening (Fig. 73),

a.

b.

Fig. 74. a. Birth process of ichthyosaur *Stenopterygius quadriscissus* (Quenst.) from the Lias of Zell near Holzmaden, Württemberg. Length 219 cm. Three young in the body cavity, the fourth just leaving its mother in typical tail-first birth. Original from Staatliches Museum für Naturkunde, Stuttgart. b. Birth of a dolphin in the Marineland Aquarium, Florida. Note similarity of position of young. (After E. J. Slijper, 1962.)

which, particularly in conical twisted shells, may extend over several whorls. Comparative trials carried out in aquariums with various marine crabs have shown that such damage to the shell is caused by hermit crabs and spiny lobsters seeking to reach and devour the soft parts. The remains of hermit crabs are seldom found in fossil form, although their life traces indicate that this scarcity is due to their unsuitability for preservation.

Traces of Reproduction

Particularly valuable are trace fossils connected with reproduction, for they provide insights into matters of physiological importance. Thanks to the complete state of preservation of ichthyosaurian skeletons from the Lias shales of Württemberg, we know that the

Fig. 75. Fossil egg clutch of dinosaur *Protoceratops* from the Cretaceous of Mongolia. Central Asian expedition of the American Museum of Natural History. Photograph courtesy of Museum of Natural History, New York.

ichthyosaurs, although reptiles, were viviparous, and we may assume that their young were born in the same manner as present-day whales, which are mammals, tail first (Fig. 74 a and b). This type of birth process may be seen as an adaptation of lung-breathing vertebrates to aquatic life.[1] It does not suggest that there is any particularly close relationship between ichthyosaurians and whales; it is an example of a convergent phenomenon.

Fossil eggs of reptiles and birds have been found in beds of various age.[2] Some of the most interesting of these are the egg clutches of dinosaurs (*Protoceratops*, Fig. 75) from the Cretaceous deposits of Outer Mongolia and the eggs of the extinct giant ostrich (*Aepyornis* etc.) of Madagascar, which attained a maximum diameter of 35 cm and contained the equivalent of 180 hen's eggs.

Living Assemblages Which Have Become Fossilized

No less interesting are fossil communities. There is a wealth of descriptions of such contemporary communities of animals and plants; for example, symbiosis, the coexistence of two partners to their mutual benefit; the case of organisms living on plants (epiphytes) or animals (epizoans) without harm to the host; and parasitism — one organism living on another and causing harm to it.

Although it is sometimes rather difficult to furnish proof of such living communities in fossils, not a few true assemblages have been described. Particularly well known among paleontologists is the life community between a Devonian coral, the tabulate *Pleurodictyum problematicum*, and a "worm" *Hicetes innexus*. The worm, which bored an S-shaped tunnel into the partly dead and partly living tissue surrounding the skeleton, has come down to us from the remains of the Lower Devonian deposits of the Schiefergebirge in the Rhineland, almost always in the form of a core. According to Schindewolf, this is

[1] Recent observations on modern whales have shown that the tail, the most mobile part of the body, is used to produce intrauterine movements and bring the fetus down into the correct position for caudal birth. All other viviparous mammals are born head first, and this again is attributable to the mobility of the head of the fetus which brings it in the uterus into the "normal" birth position.

[2] Fossil eggs are actually true fossils in the physical sense, but they are usually included among the trace fossils.

a case of "spatial parasitism." How careful one has to be in interpreting phenomena of this kind is illustrated by the coexistence of Recent corals *Heterocyathus*, etc., and sipunculids, a kind of worm of the genus *Aspidosiphon*. This relationship must be regarded as symbiotic, since when the corals are dislodged by waves in rough seas or by currents and are lying loose on the sand, the sipunculids always put them back in their proper position, which the corals themselves would not be able to do. The sipunculids give the corals the firm substrate they need for security.

Other examples include what is known as the Kerunia symbiosis, an association between hermit crabs inhabiting snail shells and *Hydractinia*, which has been described several times from Tertiary deposits; the algae living in the coats of Pleistocene giant sloths (cf. p. 108); in the Devonian psilophytes (see p. 25), the geologically oldest land plants, proof of mycorrhiza or the coexistence of fungi with the roots of higher plants; the presence of parasitic isopods in the gills of crabs (Galatheids) from the Jurassic; and the "galls" caused by parasitic worms (Myzostomids) in the stems of "sea lilies." The phenomenon of phoresia, one insect "hitching a lift" upon another, is also found in fossils. The Eocene amber contains several examples of pseudoscorpions which are firmly attached to the legs of hymenoptera (see Fig. 13, p. 22), a phenomenon which may also be observed in living insects. The clinging pseudoscorpions may be carried long distances in this way, thus extending their distribution.

Symptoms of Disease in Fossil Organisms

To mention parasitism is to take a step toward pathology, about which a great deal could be said. Healed bone fractures or damage caused by hoof kicks, for instance, have frequently been described in vertebrates. Of special interest is the proof of healed fractures or of ankylosis (coalescent bones) in the limbs of predatory animals, which gives a certain insight into their way of life. The diseases found in European cave bears *Ursus spelaeus* of the Late Ice Age — spondylosis, arthritis, periostitis, and so on — may apparently be explained by the fact that they were virtually herbivorous. This is also supported by the fact that their molars are usually ground down by constant mastication and the wear on the incisors. Such diseases are more surprising in carnivores, like the Pleistocene sabre-toothed cats from Rancho La Brea; for these diseases must have hampered movement and thus made it difficult for the cats to catch their prey. Some suspicion therefore remains that sabre-toothed cats were scavengers.

There are also various diseases which provide information about the diet of animals — for example, actinomycosis, a condition caused by blue-green algae *(Actinomyces)* which can cause the teeth to fall out and even damage the jaw by creating suppuration. Found only in grazing animals, it has been described in the fossil steppe rhinoceros and the Late Pleistocene cave bear.

The diagnosis of cases like these requires a sound knowledge of pathology and is thus better left to medical men. This is one reason why the occurrence of rickets and caries in fossil mammals, as described in the paleontological literature, is not universally accepted as fact.

VIII. Ancient Habitats

What is a Habitat?

What do we mean by "habitat"? The biologist speaks of the biotope, meaning the dwelling place of a biocenosis, thus a living community such as occurs both on the land surface and in the sea. A biotope can be mountain forest, raised bog, sandy beach, or coral reef, with all its characteristic flora and fauna. It is not, therefore, purely spatial or geographical. In paleontology it is sometimes necessary for practical reasons to employ the term "habitat" in a somewhat wider sense, and it is in this sense that we shall use it here. The reconstruction of ancient habitats, including their inhabitants, belongs to the sphere of paleoecology, as described on p. 64.

The Principle of Actualism and Its Application

In the following pages we attempt to select examples which are characteristic of paleontological habitats. We have deliberately chosen to present these examples, not in chronological order, but rather under spatial divisions, and not taking their geological age into account. The reconstruction of such habitats is attended by many pitfalls, some of which have already been described. In principle, it is important to determine whether contemporary studies can justifiably be applied in geology, in other words, whether the observation of what is happening today suffices to explain, or is even valid for, occurrences in the geological past.

Ever since there have been oceans and dry land, there have been habitats in the geographical sense in wonderful variety. On the other hand, the nature of biotopes has changed greatly in the course of Earth's history, and some biotopes that once existed have no contemporary counterpart. Starting with purely marine biotopes, we shall move by means of a series of examples to various characteristic dry-land habitats.

Let us start in the ocean depths. Are there any deep-sea fossil deposits? [1] This question is fully justified because fossils occur seldom or not at all in deep-sea deposits (abyssal sediments). Apart from the tectonic upheavals that would have been necessary for their preservation, requiring the original sea bed to have been lifted some hundreds or thousands of meters — and such upheavals are quite definitely known to have occurred — deep-sea deposits tend to be a lot less thick than shallow-sea sediments because there is little sedimentation in the deep sea, except by the turbidity currents which set up layers of sediment quite deep. It is true that there have been various descriptions of deep-sea deposits of past times, but few have stood up to critical scrutiny. With this remark we come face to face with the problems of proving deep-sea sediments. How may they be recognized? Evaluation may be based on lithological and organic components. At considerable ocean depths [2] $CaCO_3$ is dissolved, so that deep-sea deposits at depths

[1] The term "deep sea" is not uniformly defined, even in the professional literature. Deep sea is here used in the opposite sense to shallow sea, or the neritic region, which is defined as going down to 200 m. From 200 m down to a maximum of 4,000 m is the bathyal region and below that the abyssal region. Deep-sea sediments are thus beds deposited at a depth greater than 200 m, regardless of how they got there. Even the red coloration which is characteristic of many Recent deep sea deposits cannot be regarded as a safe criterion.

[2] No exact and universally valid depth data can be given, since both the solubility of the $CaCO_3$ (e. g., pteropod shells are more readily soluble than foraminiferan shells) and the degree of CO_2 saturation are important, the latter depending not only on depth but also on currents and the form of the ocean bed. It is, nevertheless, possible to state that the $CaCO_3$ content of calcareous shells is around 50% between 3,000 and 4,000 m, falling to about 20% at 5,000 m.

below 6,000 m are almost $CaCO_3$-free. Since photosynthesizing plants are unable to live at great depths because the sunlight cannot reach them, we may suppose that plants were originally absent; we know also that the composition of the animal fauna changes as the water depth increases, the profusion characteristic of shallow seas gradually diminishing. Recent deep-sea deposits consist of radiolarian and diatom mud, globigerinian and pteropodian ooze, and deep-sea red clay and sands, while the siliceous shells of radiolarians (unicellular organisms) and diatoms (siliceous algae) and/or the calcareous shells of the globigerina (unicellular organisms belonging to the Foraminifera group) and pteropods (planktonic snails) make up the essential component of the sediment. All the creatures mentioned, mostly microscopic, are planktonic organisms which float in the sea, their shells sinking to the bottom upon their death to form a sediment on the sea floor. As the depth of the sea increases, the proportion of the remains of plankton increases, while the benthic forms decrease. Thus, the ratio of planktonic to benthic inhabitants in the microfauna is an important factor in assessing the depth at which fossil marine deposits were initially laid down. However, there are various sources of error for which allowance must be made. For instance, there are various creatures living in the deep sea at the present time whose ancestors were clearly dwellers in shallow seas. Let us recall *Neopilina galatheae*, already mentioned on p. 109, and its Paleozoic relatives, the Tryblidiaceae, or the coelacanth *Latimeria chalumnae* (see p. 167), and the fossil coelacanths, e. g., "*Undina*" *penicillata* from the Solnhofen platey limestones. Neither the Tryblidiaceae nor the Mesozoic coelacanths were deep-sea dwellers. There are also deep-sea creatures which regularly make vertical expeditions, such as the lantern fish Myctophidae which come up during the night from a depth of around 800 m to the surface (and from here their bodies may be washed up in shallow-sea coastal deposits) and return to the depths at dawn. Similar movements are known in deep-sea crabs and cuttlefish. But ocean currents, too, can convey deep-sea dwellers into near-surface levels. All these possibilities have to be considered if we are to avoid erroneous conclusions. Apart from these observations, the results of submarine geology are important and have, particularly of recent years, brought about new discoveries, some of which have necessitated a revision of long-established ideas.

For these reasons, it may be understandable that in the older literature the example given here for deep-sea deposits was assigned to shallow-water seas. This is the sediment sequence found in Alpine countries and known as *flysch*, to which we have already referred on p. 120 in connection with trace fossils. The flysch sequence[3] comprises a series, sometimes several hundred meters thick, of marls, marl limestones, clay shales, and sandstones which succeed one another vertically but which can be traced horizontally over long distances. The sequence often displays a rhythmic alternation of limestone marls and graded (i. e., layered according to particle size) sandstones, which was once regarded as proof that they were formed in the vicinity of a coast. The fossils which occur in these sandstones (Foraminifera: orbitoids, nummulites; bryozoans, resins, etc., which also suggest a coastal and hence shallow-water origin) can reach great depths because of such underwater slumping as can be triggered by earthquakes. This also explains why the fossil remains always match the particle size of the sediment.

However, the trace fossils (see p. 122) and the microfauna (ratio of planktonic to benthic fauna about 90 : 10) suggested that a deeper sea must have been necessary for the formation of the sediments which — in the sandstone beds — show graded bedding, such as could have been caused by slumps and turbidity currents, and would not appear until a considerable depth below sea level, although a slight inclination of the slope of the continental shelf would be enough to achieve this. The trace fossils are exclusively those of animals which also live at greater depths on the sea floor (Fig. 76) or in the sediment (carnivorous snails, worms as sediment-feeders, etc.) and the types of trace fossils (Fig. 70) are often dependent on the composition of the sediment: e. g., helminthoids, chondrites, and "fucoids," in the limey marls; and *Scolicia*, *Palaeodictyon*, and *Spirorhaphe* in the sandstones.

The lack of features indicating a shallow-water area (raindrop impressions, wave ripples) reinforces this view. The flysch was actually

---

[3] In the Eastern Alps the name flysch is used to characterize not so much a rock type as a stratigraphic tectonic unit of the Cretaceous and Early Tertiary periods (Paleocene and Eocene). Flysch-type rocks, which indicate a flysch facies, are, however, common in the Paleozoic and Early Mesozoic eras.

deposited in deeper water and lifted and folded by tectonic phenomena to heights of a thousand or more meters above present sea level. This also explains why the flysch sediments can be traced across whole continents, and why trace fossils are abundant and macrofossils — ammonites and lamellibranchs *(Inoceramus)* — scarce.

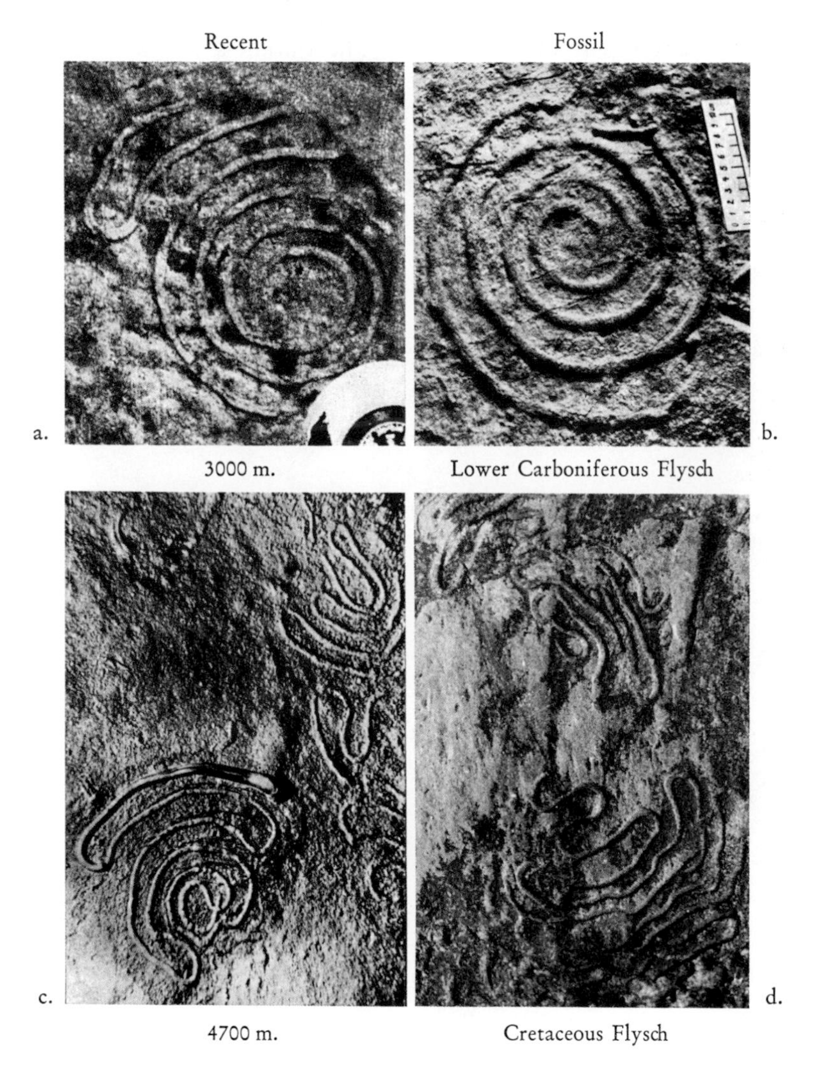

Fig. 76. Recent life traces from the deep ocean (left) and their fossil counterparts from flysch deposits (right). a. Atlantic, c. Pacific: meander track with its producer, an enteropneust. (After A. Seilacher, 1967.)

The Posidonia Shales of the Lias

In contrast, the dark Posidonia Shales known in South Germany and elsewhere are not deep-sea deposits. They are fine-grained shales, dark colored owing to the finely divided silica sulfide and bitumen, and taking their name from a tiny pelecypod *Posidonia bronni* which is the index fossil for these layers. The thickness of the Posidonia Shales fluctuates from a few meters to over 40 m. Characteristic are intercalations of limestone, usually occurring as extensive beds, and whose origin is controversial.

Their fossil content is quite different from that of the flysch. Instead of Foraminifera and ostracods, the microfossils are chiefly algae

Fig. 77. Colony of "sea lilies" *Seirocrinus subangularis* (Hill.) on driftwood overgrown with lamellibranchs. Lias from Holzmaden (Württ.) Plate 250×250 cm. Original from Senckenberg Museum, Frankfurt (Main).

and pollen grains, while there is a profusion of macrofossils. There are a few plant remains (ginkgophytes and cycadopsids) and countless lamellibranchs, snails, ammonites, belemnites, shark-like fishes, ichthyosaurs (Fig. 15), plesiosaurs, and sea crocodiles. Remains of flying reptiles, echinoids, and starfish are rare.

The area of deposition of the Posidonia Shales must have been a shelf sea. The sediment type suggests a comparison with conditions such as are presently found in the Black Sea, where, overlying toxic salt-rich deep-water layers, there is a zone of lighter, salt-poor water which prevents aeration of the bottom. However, this conclusion does not quite fit the case: the Posidonia Shales were found — as may be deduced from the occurrence and distribution of the fauna, not to mention the worm burrows *Chondrites bollensis* in the sediment — under both gyttja and sapropel slime conditions; the first were partly rotted and the second completely rotted, which means that life was possible on the bottom for only some of the time. The lack of oxygen, while preventing any abundant sea floor life, encouraged the generally excellent preservation of the fossils — the organisms did not decompose before burial (see Fig. 16 and p. 9). The sessile "sea lilies" did not, however, grow on the sea floor itself but generally on driftwood (Fig. 77) which sank to the bottom and was covered by sediment. Much the same applies to the pelecypods *(Posidonia,* etc.) which were attached by their byssus (a tuft of filaments for clinging) to drifting seaweed or wood. The harpoceratids (ammonites), too, are regarded as dwellers in seaweed.

The Dachstein Reef Limestones

The Dachstein reef limestones derive from a totally different habitat; they form often very thick mountain massifs in the northern (Salzkammergut, Austria) and southern (Dolomites of South Tyrol) limestone mountains of the Alps. Such a mountain range, consisting mainly of reef limestones, is the Gosau Ridge (Fig. 78) to the west of the Dachstein, whose interesting morphological variety makes it a mecca for climbers. Although fossils are abundant in these reef limestones, it is not usually easy to work out their systematic position. This is due less to the tectonic stressing of the rock than to the types of organism which built up the reef limestone. Comparative studies of such fossil reef limestones have quite recently been begun again very intensively in order to investigate the sedimentological composition, or microfacies; Recent atolls also have received close attention only in the last few decades when their structure has been studied from cores (Eniwetok and Bikini atolls in the Pacific, for example). These

Fig. 78. The Gosau ridge near Gosau (Upper Austria). Typical mountain range of unbedded, vertically cleft Dachstein reef limestone of the Upper Triassic. In foreground karst formation in Dachstein limestone. (After F. Simony from H. Zapfe, 1957.)

studies have substantially increased our knowledge of the structure of these former reefs, [4] although the foundations were laid over 100 years ago by the work of F. von Richthofen and E. von Mojsisovics. Coral was once thought to be mainly responsible for these Upper Triassic reef limestones, which were assumed to be uniform reef bodies. Today we know that in addition to the coral (stone coral or scleractinians:

---

[4] The word "reef" tends to be replaced in the modern literature by "bioherm" insofar as these reefs have been built up by organisms.

in particular, *Thecosmilia,* the Dachstein coral, and the Thamnastereids, Montlivalteids, and Stylophyllids) the main *reef-building* organisms are calcareous sponges (Sphinctozoa), calcareous algae (Codiaceae, Dasycladaceae, Solenoporaceae), hydrozoans (so-called Spongiomorpha), and bryozoans (Fig. 79). Mussels, snails, sea urchins, crabs, and fish, on the other hand, were simply *reef dwellers.* The reef body itself is not a massive structure but is composed of a multitude of separate reefs with the interstices filled up with coral detritus and shell debris, produced by the action of the waves on the reef-dwelling organisms and washed into the fissures. The "cement" is mostly

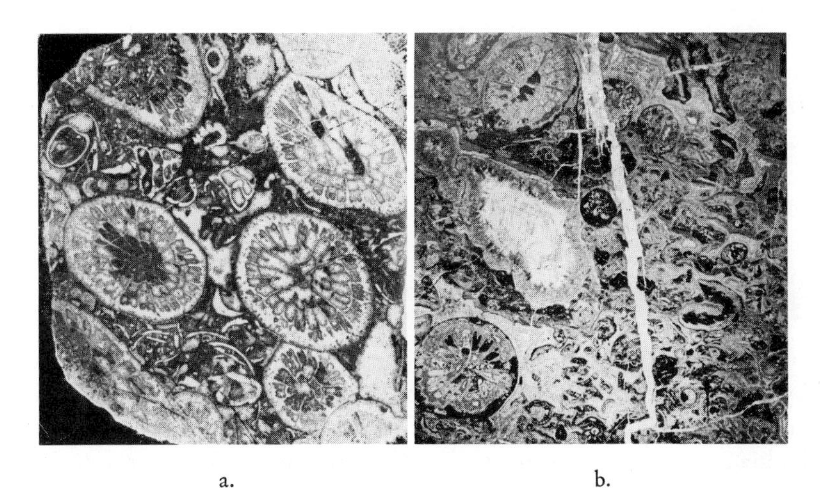

a. b.

Fig. 79. Thin sections of typical Rhaetian reef limestone from the Rötelwand, near Hallein, Salzburg. Corals *Thecosmilia clathrata* (Emmrich), sponges, stony green algae (Dasycladaceae: *Macroporella* etc.) and small gastropods. a. Reef debris not transported far. b. Area of reef slope, about natural size. (Photos courtesy of E. Flügel, Darmstadt.)

provided by calcareous algae which "fix" the reef and weld it into a whole. This again means that these reefs could have originated only in shallow waters, as is known from studies of Recent coral reefs. The enormous thickness of the Dachstein reef limestones may be explained by constant sinking of the sea floor, as has been attested for many marine basins throughout geological history. The reefs themselves were the habitat of numerous fish which have much in common with Recent coral fishes (e. g., full-bodied and having pebbly teeth) but without being closely related to them. For in those days there were no "true" coral fish — they were nearly all ganoid fishes, many of whose Recent relatives are fresh-water fish. In addition, primitive sharks and marine reptiles were common. Among the latter, the Placodontia were certainly true reef-dwellers. Many of these resembled sea turtles in their habitus (e. g., *Placochelys),* their limbs being transformed into flippers and their bodies protected by a rigid armor of bone and horny plates. Their dentition consists of a few enlarged single teeth, which look like pebbles and must have served to crush molluscs and brachiopods. Although no clues as to the climate in those days can be obtained from the fish and marine reptiles, both the coral reefs [5] and the various kinds of algae suggest that the conditions under which these reefs originated were tropical. The highest peak in the Dachstein (Hoher Dachstein) consists of bedded Dachstein Limestone, deposited in the lagoons of the backreef area.

Our next example will be a shallow sea which formed the habitat of countless vertebrates.

The Niobrara Sea of the Great Plains

The Midwest of the North American continent was once covered by boundless prairie grassland, now generally under agricultural cultivation. This prairie was first formed in the Tertiary period and its lower strata today consist of shallow nonmarine Tertiary deposits. As late as the Cretaceous period, large parts of the Great Plains were covered by a sea which filled a bay extending from what is today the Gulf of Mexico as far as Canada. Among the special features of the deposits of this shallow sea are the Niobrara limestones, whose bizarre erosion shapes form a striking feature of the landscape. These are chiefly found on the edge of the Central Great Plains, from Kansas via Colorado and Nebraska to North and South Dakota. The thickness of the Niobrara formation fluctuates from the south (Bighorn Mountains) to the north between 230 and 30 m. There are littoral

[5] Recent reef coral thrive only in warm, shallow seas where the temperature seldom falls below 20° C, i. e., mostly in the tropics. Further, reef corals grow only in water where sunlight can penetrate and never deeper than 50 m. Their light dependence is confirmed by the fact that algae (Zooxanthelles) live symbiotically in the coral. There is no proof that Zooxanthelles actually did exist in the Triassic coral, but it seems very likely.

limestones from the Late Cretaceous period (Niobrarian = Lower Senonian) which have long been famous for their prolific fossil content. They bear witness to the great variety of life which once thrived in this huge bay. Apart from a few remains of land plants and land animals which were washed down by floods, these are all marine animals.

Particularly characteristic are colonies of inoceramids (*"Haploscapha" grandis*) and whole banks of oysters of the small *Ostrea congesta* species. These provide proof that this shallow sea was not too distant from the coast. Fossils are still being exposed by weathering, particularly the remains of fishes, reptiles, and birds. These occurrences have been known for a long time. As early as 1875, E. D. Cope described 48 species of fish, 37 species of reptiles, and 4 species of birds from the Niobrara Limestone; collections of thousands of bones made since then have added to these. Worthy of special mention are the remains of the mosasaurs — *Platecarpus, Tylosaurus, Clidastes;* the plesiosaurs — *Elasmosaurus* and *Trinacromerum;* and the giant flying reptiles of the genus *Pteranodon,* long since extinct.

The mosasaurs, named after the Maas River near Maastricht (Belgium), the place where they were first discovered, are members of the order of Squamata, "scaly" reptiles, and hence giant relatives of today's agamids and monitor lizards, which had adapted to life in the water. These giant lizards had an elongated body some 5–10 m long, and their limbs were transformed into flippers. Their tail was long and flattened at the sides and, like the crocodile's, used for propulsion. To judge by its long fangs, the mosasaur lived on fish, cuttlefish,[6] turtles, and sometimes other marine animals. A special feature they possessed was an extra joint in the middle of the lower jaw which enabled the jaw to be extended downward so that large animals could be swallowed whole. Mosasaurs are so far known to have existed only during the Cretaceous period, when they enjoyed a worldwide distribution.

The plesiosaurs, on the other hand, are known to have existed in the Jurassic period as well. The most remarkable genus is undoubtedly *Elasmosaurus,* a giant form whose neck accounted for some 7 m of its overall length of 13 m. It had a small head, limbs fashioned as slender flippers, and a short tail which cannot have been much help in pro-

---

[6] Ammonite shells have been found bearing the imprint of mosasaur teeth.

Fig. 80. Reconstruction of the Niobrara Sea in Kansas with *Tylosaurus* (Mosasauria) in the foreground, the long-necked plesiosaur *Elasmosaurus* and the pterosaur *Pteranodon.* (After J. Augusta and Z. Burian, 1956.)

pulsion. *Trinacromerum* had a long head and a short neck, but both genera were fish-eaters.

In addition to the mosasaurs and plesiosaurs, there was a third group of reptiles inhabiting the Niobrara sea: the turtles. There are numerous remains of swimming turtles (e. g., *Toxochelys, Protostega),* recognized by their armored shells and limbs. Both the dorsal and ventral shells are flat and exhibit the fontanelles, typical of marine turtles, which represented a saving in weight.

Above the sea hovered creatures rather like the albatross, the giant flying reptiles *Pteranodon ingens.* These toothless, fish-eating, flying reptiles had a wing span of 7 m, exceeding that of the largest birds. The flying reptiles had an extremely elongated fourth digit which served to extend the flying skin. The three forefingers had claws and were freely movable.

No less remarkable was one of the birds of that period, *Hesperornis regalis,* a large flightless, long-necked sea bird whose habitus is reminiscent of the modern grebe *Podiceps.* The powerful back feet were used as oars, the forelimbs being mere rudiments of the upper arm bones. The resemblance to the grebe is simply a convergence phenomenon, as is clear from the structure of the skull and the toothed jaw. *Hesperornis regalis* was also a fish-eater and is presumed to have lived in large colonies along the shores of the Niobrara sea.

There were also numerous species of fish, including sharks — *Galeocerdo, Oxyrhina,* and *Lamna;* rays — *Ptychodus*; ganoid fishes — *Protosphyraena;* and generally primitive bony fishes — teleosts *"Portheus," Pachyrhizodus, Ichthyodectes, Saurocephalus,* and *Saurodon.*

The rays had a true pebbly dentition, adapted to cracking open lamellibranchs and crabs. The bony fishes included giant forms over 5 m long, such as *"Portheus" (Xiphactinus) molossus.*

Remains of ammonites *(Scaphites)* and free-swimming crinoids *(Uintacrinus socialis),* whose calyces often cover large areas of bedding planes, indicate that there was a clear way to the open sea and that currents sometimes beached hundreds and thousands of these creatures.

We shall now leave the sea and pass on to fresh-water forms and purely terrestrial habitats.

The Oeningen Molasse

The beds of the upper fresh-water molasse of Oeningen on Lake Constance are typical nonmarine sediments. These deposits have so far yielded more than 1,500 fossil remains of plants and animals. In addition to fresh-water animals, they contain a great abundance of insects, reptiles, birds, and mammals, as well as land plants, thus giving some idea of what living creatures were like at that time. These are deposits of the late Miocene age, composed of a succession of sands, rubbles, marls, fresh-water limestones, clays, shales, coals, and volcanic products (tuffs and tuffite, that is, water-worn and sedimented tuffs). Its wealth of fossils early attracted the attention of collectors and naturalists. On account of the "Homo diluvii testis" (see p. 5), Oeningen became famous far and wide. This specimen, in fact of the giant salamander *Andrias scheuchzeri,* turned out to be the best-known species among the skeletal remains of Oeningen. Furthermore, in the nineteenth century the pioneer studies of Swiss naturalist Oswald Heer made the Oeningen fossil locations on the Schienerberg into classics of paleontology. Heer ascribed the differences in the fossil content of the various beds to seasonal changes.

The landscape must have consisted of lakes and rivers with many stagnant pools. Some of the pools were poorly aerated and, because of the rotting of large amounts of plant debris, formed a medium very poor in oxygen. This was the habitat of many aquatic plants — hornweed, water fern *(Salvinia),* and reed mace; of snails — *Planorbarius, Radix, Ancylus,* and *Theodoxus;* of fresh-water mussels — *Unio lavateri, Margaritana flabellata,* and *Pisidium escheri;* of crabs — ostracods and crayfish; of various insects — dragonflies, thread-horn larvae, and water beetles; of fresh-water fish — whitefish and pike; of giant salamanders and toads *(Latonia);* of giant crocodiles and turtles — the latter of the swamp *(Emys),* snap *(Chelydra)* and soft-shelled *(Trionyx)* types; and of assorted water birds and of beavers.

The mixed woods, which came down to the water's edge and were flooded for part of the time, consisted of reeds *(Phragmites),* willows and alders, water elms *(Zelkova),* hickory *(Pterocarya),* plane, maple, elm, sweet gum *(Liquidambar),* and water pines *(Glyptostrobus).* The dry hinterland was covered with steppe grasses, so-called hardwood mesophytic plants and tree-like shrubs *(Podogonium, Robinia, Gleditschia,* etc.). Apart from the mixed woods and "tree savannah," there were clusters of palms and laurel woods with cinnamon or camphor trees *(Cinnamomum), Sassafras, Persea, Myrica, Sapindus, Aralia, Diospyros,* etc. This vegetation, taken as a whole, suggests a warm-temperate to Mediterranean climate.

Apart from the aquatic plants, there are not many remains of the herbaceous plants to be found. It is surprising what an abundance there was of woody plants, far more than is the case in Central Europe today. This was due to the climate at that time, as was also the occurrence of plant genera today restricted to Asia or North America, such as *Liquidambar, Sassafras, Cinnamomum, Zelkova, Pterocarya, Persea, Sapindus, Glyptostrobus,* etc. The last are often grown nowadays in parks in Central Europe, which shows that it is not the climate that accounts for their absence among the natural Central European flora. The true cause of their absence was the Ice Age, which followed the Tertiary period and brought about the disappearance of elements that were susceptible to climate. Because of the barrier formed by the Alps and the Mediterranean Sea, these plants were unable to move southwards, as they did in North America or East Asia where the mountain ranges run from north to south.

The wooded landscape itself was the home of an enormously rich world of insects, as is attested chiefly by the remains in the insect-shales, the "Book of Nature," as Heer called it. The woods also sheltered the elephant-like mastodon *Bunolophodon angustidens,* rhinoceroses, short-necked giraffids *Palaeomeryx eminens,* wild pigs, carnivorous mammals, squirrels and dormice. The open landscape was inhabited by land turtles and rodents resembling hamsters and pikas.

Coal

We have been describing a lake and river landscape together with its environment and its flora and fauna, which were certainly rather different from those of today. Our next example will be the typical coal swamps, which again were quite different from present-day bogs, and not in their flora alone. As the coal basins of Rhineland-Westphalia, England, Pennsylvania, and Illinois contain a great wealth of plant fossils, let us begin with these areas.

In present-day land vegetation, angiosperms and gymnosperms always represent a much larger proportion than the sporangia-bearing plants — pteridophytes; and ferns, club mosses, and horsetails — mosses (bryophytes) and thallophytes (algae, lichens, and fungi). During the Carboniferous period the pteridophytes were the most abundant plants. They are also the most important coal-building plants, and some of them were indeed arborescent and not, like their living relatives (club mosses and horsetails), herbaceous plants. Consequently, we must picture a strange-looking type of Carboniferous forest (Fig. 43, p. 62). After many fossil discoveries and years of intensive research, it is now possible to make exact reconstructions of these coal-forming plants.

The most important coal-forming plants were the lepidophytes, which were abundant in both species and number, the best known being the scaly trees *(Lepidodendron)* and *Sigillaria*. Total heights of over 30 m with a bole diameter of over 2 m were no rarity. The roots and branches of the lepidodendrons always forked a number of times (dichotomous). The branches were covered by long lanceolate leaves which fell off to leave the characteristic scars which give the tree its name. The sigillarids were unbranched or simply forked at the top of the shoot and bore stiff, grass-like leaves which formed tufts, so that they are also called tuft trees. The sporangia grew straight out of the trunk just below the leaf tuft, a phenomenon which corresponds to the cauliflory (bearing flowers on the trunk) of many tropical trees — the cacao tree, for example.

The horsetails of the Carboniferous were also arborescent plants, which is why they are called giant horsetails *(Calamites)*. They were similar in structure to the horsetails *(Equisetum)* of today, and they attained an overall height of nearly 20 m with a bole diameter of almost 1 m. Other tree-like plants of the coal swamps were tree ferns and cordaitids, which are true gymnosperms. True ferns and the seed-bearing ferns that resembled them (pteridospermatophytes) to some extent took the place of bushes, but there was no lack of real herbaceous plants: there were some among both the lycopods and the mosses. At that time, neither conifers nor angiospermous woods existed, nor did flowers. The role of liana-type creepers was played by sphenophyllids (wedge-leaved plants), thus filling an ecological niche.

Although we can do no more than speculate about temperature conditions at that time, temperature and rainfall must certainly have been relatively even all the year round. The chief indicator of this is that there were no dormant buds. [7]

The coal swamps formed the habitat of various small crustaceans; fresh-water mussels and fish, including lung fish; and amphibians

---

[7] The lack of growth zones (annual rings) in woody plants does not, however, justify the conclusion that growth was uninterrupted and hence that the climate never varied at all.

(labyrinthodonts and lepospondyls) whose main representatives resembled salamanders and snakes. Other inhabitants of the coal forests were ancient forms of spiders, centipedes, and insects, some individuals reaching vast proportions — for instance, *Meganeura,* with a 75-cm wing span. There were primitive dragonflies, cockroaches, mayflies, and orthopterids, but "modern" insects like beetles, bugs and butterflies, and bees were entirely lacking. They simply did not exist at that time.

Amber

After our venture into the Carboniferous period, we now return to the Tertiary period and discuss the Scandinavian amber forests of the Early Tertiary period, which are very well known to us from the shores of the Baltic (U.S.S.R., formerly East Prussia), whose amber was studied first by Breslau paleontologist H. R. Goeppert, and his pupil H. Conwentz.

In this case, too, we have fossil remains which have been known to men for thousands of years; amber was already known to Late Paleolithic man and was used for ornament and worn in the form of pendants by Middle and Late Stone Age man. During the Bronze Age amber from the north found its way to the peoples of the Mediterranean (it is Pliny's "succinum"). At certain periods there were true trade routes whose course has been reconstructed partly from excavations and partly from the writings of the ancient authors. Naturally, other raw materials were carried and traded along these routes.

Amber is the resinous exudation of the fossil amber pine *Pinus succinifera,* and it is graded in the trade according to its clarity and its ability to take a polish as clear, smooth amber, or bastard, bone or foam amber. The Baltic amber — and this is all we shall discuss here — which is found mostly in irregular nuggets, drops, or dish-shapes, occurs chiefly in Early Tertiary (Oligocene) beds, and particularly in what is called "blue earth," glauconitic sands in secondary deposits on the coast of the Baltic. Here the amber-bearing sediments are eroded by the waters of the Baltic, and it is here that it is mined in commercial quantities or fished up in nets in seaweed wrack after storms. However, in addition to its commercial importance in the jewelry trade, as an insulating material, and in the manufacture of amber lacquer or succinic acid, amber constitutes for the paleontologist, on account of its inclusions of animals and plants, one of the most valuable records of the past. We have already mentioned the often immaculate state of preservation of these inclusions (Fig. 13, p. 22). Above all, they give an insight into the small animal population of the amber forests and give some pointers as to the composition of the vegetation. To be sure, the amber inclusions which have come down to us in their entirety as so-called oryctocenoses represent a mere cross-section of the primitive profusion of types of organism, but even this enables us to make many deductions about what the habitat was like at that time.[8] More recent investigations have shown that the amber forests were essentially mixed pine forests, associated predominantly with cypress (Cupressaceae and palms, also oaks and chestnuts. Such species composition is known today in the subtropics and tropics of the southeastern part of North America and the Antilles. However, these pine forests do not exactly deserve to be called "forest", in that "forest" may be regarded as a closed complex. There were several types of mixed pinewood with clearings and larger open spaces covered with heather (Ericaceae), and there was no shortage of running water. Another characteristic feature was that swamp cypresses (Taxodiaceae) had completely disappeared. This association of plants suggests a warm, dry climate, and the insect population tends to confirm this. Among the amber arthropods there were some forest dwellers with definite thermophilic elements — termites, paussids, and so on. Species that lived in bark — woodlice, bark beetles, stamping beetles, and some musk beetles — were numerous, as were species that lived in the soil on fallen leaves, pine needles, and rotten wood — springtails, centipedes, mites, woodlice, scorpions, earwigs, cockroaches, etc. But even fresh-water insects — such as flea crabs, mayflies, caddis fly larvae, water beetles, water boatmen, and threadhorn larvae — were not lacking. Nevertheless, this is simply a selection, determined by the potential for preservation offered by amber; thus, there are virtually no remains of larger animals which did not stick to the resin or could not be engulfed by it. Only the hair of mammals and feathers of birds are found.[9]

---

[8] Supposed marine fossils found in amber (corals: *Hydraulis)* are pseudofossils, due to ageing processes in the amber.

[9] Frogs and lizards in amber are almost without exception counterfeit, apart from the various confirmed examples of lizard exuvia.

The Tar Pools of Rancho La Brea

The asphalt swamps of Rancho La Brea in the vicinity of the present city of Los Angeles were first mentioned in 1769 by the Spaniard Gaspar de Portola and are known to paleontologists the world over as the largest collection of bones anywhere (see Fig. 11, p. 21). The tar pools which were once so common there are simply surface escapes of oil. Most of them have now vanished, the last ones being protected within a National Park; the fossils are displayed in a museum.

Petroleum and natural gas tend to migrate upward in porous sediments or via cracks in rock on account of their specific gravity. As a result of "weathering," the petroleum on the surface forms asphalt swamps, which present a tough layer of tar. The surface of the sticky tar pools is usually covered with rainwater, which makes them real animal traps, as mentioned earlier. These tar pools have been in existence since the Late Pleistocene age, and since then countless creatures have been attracted to them and met their end, held fast in the sticky tar (see Fig. 12, p. 22). These creatures in turn attracted predators, both birds and animals. This fact is clear from the composition of the fauna, the number of carnivores exceeding that of the herbivores by a factor of ten. According to C. Stock, birds of prey make up more than 50% of the birds found there.

Even in the last century, little attention was paid to the enormous quantities of bones and teeth which were found during the extraction of the asphalt. Only in 1905 was their fossil nature recognized. Scientific excavations started in that year by the University of California under the direction of Prof. J. C. Merriam led to the recovery of thousands of skeletal remains of Ice Age vertebrates.

Although the skeletal remains from the tar pools of Rancho La Brea represent a limited cross-section of the vertebrate fauna of the Late Pleistocene age, the completeness of the skeletons thus preserved gives us a most valuable picture of the animal world as it existed at that time.

Of particular interest are the proboscideans, which have been extinct in North and South America for thousands of years. These include elephants *Archidiskodon imperator*, mastodons *Mastodon americanus*, and other mammals: giant sloths *Paramylodon* and *Nothrotherium*, camels *Camelops hesternus*, one-toed horses *Equus occidentalis,* short-faced bears *Arctodus simus* and sabre-toothed cats *Smilodon californicus.* Predominant among the birds were the diurnal birds of prey, including a huge condor-like vulture, *Teratornis merriami* being the most striking species, then the turkey vulture *Cathartes aura,* which is still an inhabitant of California today, condors *Gymnops amplus* and *G. californicus,* vultures *Neophrontops americanus* and *Neogyps errans,* resembling to some extent the old-world carrion vultures, also hawks, buzzards, and eagles *Aquila chrysaetos.* The remaining birds do not include many species that have become extinct, although there are the Californian turkey *Parapavo californicus,* the tar stork *Ciconia maltha,* and the La Brea owl *Strix brea.* Also found in large numbers are the remains of chicken and geese, cranes, herons, divers, dotterels, woodpeckers, owls, doves, cuckoos, and starlings. On the other hand, remains that are not found so often include swamp turtles, toads, fresh-water snails *Lymnaea,* centipedes *Spirobolus,* spiders and insects (termites, locusts, cicadas, flies, beetles, wasps, and ants) and plants — pine *Pinus muricata,* cypress, juniper oak *Quercus agrifolia,* elder *Sambucus,* and hackberry *Celtis mississippiensis.*

The fossil flora and fauna indicate that an open landscape, with patches of bushes here and there, must have surrounded the tar pools and that there was woodland along the banks of small rivers. The climate at that time must have been more humid than at present, with similar or slightly higher temperatures and a rather higher annual rainfall than today.

Why are so many animals now extinct when present living conditions are not apparently basically different from what they were at that time? Extinction occurred at the end of the Ice Age or even in the Early Holocene and must have been connected with a fluctuation in climate which brought about a constriction of the habitat.

The late Tertiary amber forests and the Quaternary tar pools are representative of mainly terrestrial habitats. Our last example of a fossil habitat will be of a purely terrestrial biotope. It is the Late Pleistocene loess steppe of Central Europe, where the flora and fauna have been well preserved in fossil form.

The Central European Loess

The Late Pleistocene loess, which is very extensive — particularly in the northern part of Lower Austria, in Czechoslovakia, Hungary, the Southern U.S.S.R., and so on — lends a most characteristic appearance to many districts. The loess is a yellowish to light gray, porous, unbedded fine sand which was for the most part brought by the wind from the huge outwash plains and moraines and deposited principally on the southern and eastern slopes of the hills, which were at that time unforested. This fact itself — lack of closed tree cover — suggests that the deposits occurred during a cold period, an assumption which is strengthened by the discovery of Arctic and Alpine fauna, also by the so-called loam or clayey zones, which are in fact soils formed during warm periods, distinguished from the loess by their color, brownish to reddish according to the duration and intensity of the warm period. Thus, the loess profiles reflect the several fluctuations of cold and warm periods known to have occurred during the Pleistocene age. During the cold periods, the tree limit retreated by about 1,000–1,200 m compared with today, or moved further south, because of the fall of 8–12° C in the average annual temperature. Large areas of Central Europe were thus unforested during the last cold period. Fossils have shown that only dwarf bushes — dwarf willows or birches — were able to survive; these bushes, together with dwarf pines in the foothills of the Alps and tundra plants (e. g., *Dryas octopetala)*, made up the vegetation of this hilly landscape, the home of a fauna whose remains are found in the loess. As regards the flora, we shall merely say that the edelweiss *Leontopodium alpinum,* today regarded as the epitome of Alpine flora, was originally a plant of the steppe, as indeed it still is today over extensive areas of Asia. Whereas the flora of that time essentially differed from today's only in its spatial distribution, the Late Pleistocene loess steppe fauna was largely made up of species that are now extinct. Many surviving forms, like the plants, are found today only in the Alps, northern Europe, and the Arctic, or in the steppes of Asia.

The characteristic large animals of the loess steppes include the mammoth *Mammuthus primigenius,* the woolly rhinoceros *Coelodonta antiquitatis,* the steppe bison *Bison priscus,* the giant deer *Megaceros giganteus,* the wild horse *Equus "germanicus,"* and the wild ass *Asinus hydruntinus,* together with the cave hyena *Crocuta spelaea,* all of which are now extinct. Other characteristic fauna of the loess were reindeer *Rangifer tarandus,* ibex *Capra ibex,* musk ox *Ovibos moschatus,* arctic fox *Alopex lagopus,* wolf *Canis lupus,* brown bear *Ursus arctos,* wolverine *Gulo gulo,* Arctic hare *Lepus timidus,* marmot *Marmota marmota,* white grouse *Lagopus,* and lemmings *Dicrostonyx.* But there were also chamois *Rupicapra rupicapra,* Saiga antelope *Saiga tatarica,* pika *Ochotona pusilla,* jerboa *Allactaga jaculus,* polecat *Putorius putorius eversmanni,* and birch mouse *Sicista montana.* Of the invertebrates, only land snails have been found in the loess, and these are valuable ecological indicators.

The paleogeographer is directly interested in the reconstruction of biotopes of the past and therefore in the facies. From fossils he can not only assess the ancient distribution of land and sea and the climate of the time, but can also draw certain conclusions about the bridges of land between continents.

IX. Fossils and Paleogeography

The question of the paleogeography of the times is often tied up with facies. This question crops up not only in such matters as the evaluation of the flysch facies as a sediment deposited in a deep-sea trough, but also in smaller areas involving such concepts as reefs and lagoons, as mentioned in the discussion of the Dachstein limestones of the Upper Triassic.

Land Bridges or Continental Drift?

The frequently extensive deposits of what were once marine sediments on the surface of the various continents, the emergence or disappearance in recent times of volcanic islands, the drying up of inland seas and similar phenomena justify the supposition that the pattern of the Earth's surface was similarly subject to constant change in the past. Add to this the modern pattern of the distribution of many plants and animals which is hard to explain on the basis of the present position of oceans and continents, not to mention the findings of geology, paleomagnetism, and oceanography.

Paleogeography deals with the geographical phenomena of early times, ranging from climate, by way of the conformation of the Earth's surface, to such matters as the constancy of oceans and land bridges, or continental drift.

Quite early on, the botanical and zoological geographers recognized that there were links between the flora and fauna of continents which are entirely separate today, or between those of continents and islands, such as South America and New Zealand, Africa and South America, Borneo and Southeast Asia, Australia and New Guinea, and

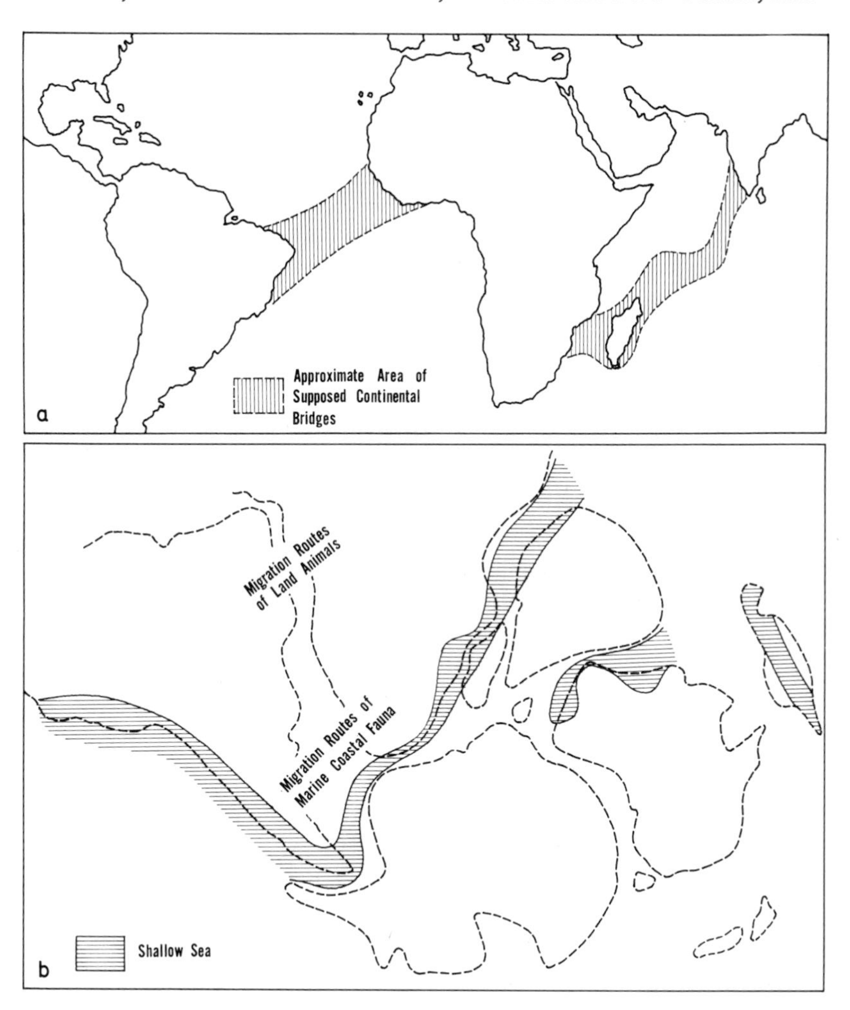

Fig. 81. Land bridges or continental drift? a. after Ch. Schuchert, 1932; b. Gondwanaland in the Mesozoic after A. Hallam, 1967. Continental drift or ocean-floor spreading now fully confirmed by geophysics (paleomagnetism), geology, paleontology, and oceanography. Timing of drift and path followed by the various continental masses still controversial.

also between oceans which are today divided by land bridges, e. g., the Eastern Pacific and the Caribbean Sea. Although it might appear relatively simple to explain the latter by the existence at some earlier time of connections between oceans (the Panama Road and Bolivar syncline in Colombia), or the occurrence of many large land animals on Borneo or New Guinea by a slight drop in the sea level (see below), such explanations completely fail to account for the other phenomena. Even the assumption that the flora and fauna in question once had a worldwide distribution provides only a partial explanation. Had there once been land bridges? Or had there been direct contact, broken by the drifting of continents? For decades the hypothesis of continental drift, put forward chiefly by A. Wegener in 1912, was hotly contested by geologists and geophysicists, but during the last ten years the dissenters have been coming around to Wegener's way of thinking, mainly as a result of paleomagnetic measurements (Fig. 81). Paleomagnetism is based upon the discovery of remanent magnetism, i. e., the iron particles in magmatic or sedimentary rocks — e. g., magnetite, hematite and other Fe minerals — are oriented in a certain direction on the basis of the magnetic field which existed in the geological past. From this, conclusions may be drawn about the position of the magnetic poles at the time when this "fossil" magnetism was active.

Geological, paleontological, and oceanographic findings since then have not only confirmed in principle that continental drift did occur, but they have also supplied exact data about the nature of the original contact and even the time at which the land masses parted. Even the causes of continental drift may be considered to have been explained in principle by the existence of convection currents under the "rigid" crust of the Earth. [1]

The discovery of midocean ridges in the Atlantic, Indian, and Pacific Oceans, and the fact that the sediments forming the ocean floor gradually become younger the nearer they are to the ridge, are further clues to what R. S. Dietz has called "ocean-floor spreading" and also to the direction and motive force of the drift.

Let us give just one example which shows how fossils play an important role. The astonishingly good fit between the edges of the continents of Africa and South America was the decisive point for

---

[1] It is not possible within the scope of this book to discuss the problem, which has recently become topical again, of the expansion of the Earth and the significance of this continental drift.

Wegener when he developed his theory of continental drift. Recent oceanographic investigations have shown, however, that it is not the present coastlines but the edges of the continental shelves which make an almost perfect fit.[2] In addition, geological studies have confirmed that the rock series match on the east coast of South America and the west coast of Africa. These and the long-recognized traces in both continents of glaciation from the Late Paleozoic era can only be explained by direct contact. South America — or the Brazilian Shield, as it is more appropriately called — and Africa were once part of one great southern continent which the Viennese geologist E. Suess, as early as 1875, named Gondwanaland after a part of India, although he assumed that there had been land bridges. This southern continent had originally comprised Madagascar, India, Australia, and the Antarctic continent as well. Wegener assumed there had originally been a single continent, Pangaea, but the present theory is that that was a northern continent, Laurasia, and a southern one, Gondwanaland, and that these were separated by the Tethys.

We shall discuss only a small part of the over-all problem. For how long did direct contact persist between South America and Africa? Or put another way: When did the South Atlantic Ocean form? Quite recently K. Krömmelbein identified largely similar microfaunas (chiefly ostracods) from profiles of nonmarine basin sediments which he found in microsamples from cores taken in East Brazil (Bahia Province) and West Africa (Gabon) while prospecting for petroleum. These faunal successions provide proof that even in the Early Cretaceous (Neocomian) period there was a common inland sea which covered parts of East Brazil and West Africa. It was not until the end of the Early Cretaceous period that the two continental masses began to move apart, thus forming the South Atlantic. This information obtained from the study of microfossils has since been confirmed by oceanographic investigations which showed that the floor of the South Atlantic had no sediments earlier than the Cretaceous period. Hence, the splitting of Africa and South America must have begun during the Cretaceous period. This finding is supported not only by the distribution at that time of coast-dwelling molluscs, which have been found from East Africa to Argentina, but also by numerous observa-

[2] Three places where overlapping occurs—near the mouths of the Niger and Congo Rivers and in Walvis Bay—result either from the formation of deltas since the separation of the continents, or from tectonic upheavals of the sea floor.

tions on plant and animal geography which would be difficult or impossible to explain other than by direct contact persisting right up to the Cretaceous period. We shall mention just a few samples here.

Disjunctive Distribution of Recent Organisms

The evaluation of zoogeographical observations in order to reach a decision about former land contacts is possible only in certain circumstances. In the case of terrestrial animals, one must first exclude passive carriage by wind or ocean currents, or possible flight (e. g., copepods introduced by migrant birds) at least as firmly as when establishing a monophyletic origin. According to W. Hennig, only parallel groups which can be traced back to a common root group and which occur disjunctively at the present time on two different continents can support the conclusion that there was once contact between the areas of distribution. Furthermore, the geological age of the animal group is important, and this again may be evaluated from fossils.

Frequently, only geologically old groups of animals show signs of close relationship, while this does not apply in the case of the geologically younger groups. From South America we will mention among the geologically old groups various fresh-water fishes (lung fish: *Lepidosiren;* osteoglossids: *Osteoglossus* and *Scleropages;* and characins: Characidae); amphibians (Pipidae: *Pipa);* reptiles (pelomedusid turtles: *Podocnemis);* and the rhea *(Rhea).* Their closest relatives (e. g., lung fish: *Protopterus;* osteoglossids: *Heterotis;* characins: Hydrocynidae, Citharinidae, etc.; Pipidae: *Xenopus;* Pelomedusidae: *Pelomedusa;* ostrich: *Struthio)* all live in Africa now (Fig. 82). They are all members of geologically old groups, as documented by the fossil record. In contrast, there are very great differences between animal groups which had their origin in the Holocene age. Where resemblances do exist, they are not due to the fact that the animals are related but rather to parallel and convergence phenomena. Thus, the porcupine-type rodents long regarded as very closely related — Caviomorpha in South America and Hystricomorpha in Africa — and the monkeys — New World monkeys in South America and Old World simian primates in Africa — turn out to be members of parallel lineages which, despite their often amazing likenesses e. g., new-world tree porcupines and old-world porcupines; new-world spider monkeys and old-world colobus monkey; are not directly related. There are various other organisms displaying typical convergence phenomena, the best-known of these being the new-world humming birds, iguanids and cacti and their old-world counterparts, nectar birds, agamids, and succulents (Euphorbiaceae).

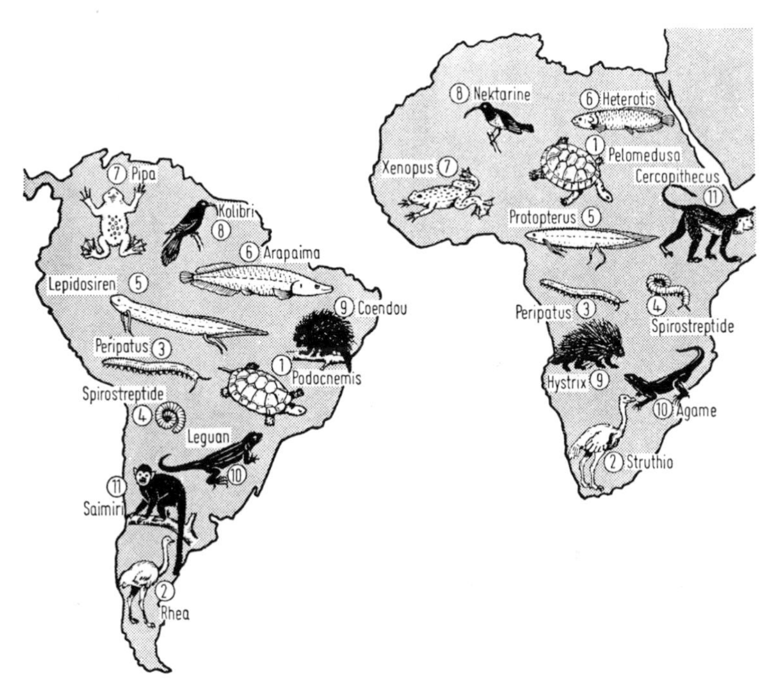

Fig. 82. Animal geography and continental drift. Geologically old (light) and young (dark) animal groups of South America and Africa. Note agreement between geologically old forms—close relatives or identical genera. Resemblances between geologically young animals due to convergence. (From E. Thenius, 1970.)

As already stated, the present disjunctive distribution of animals and plants may point to a once-continuous area of distribution. The many examples confirmed by fossils include, among mammals, tapirs, elephants, rhinoceros, and the various camel types (camels and llamas); among flowering plants, the *Araucaria* and southern beech *Nothofagus.*

Tapirs are today restricted to Central and South America and South Eastern Asia. Fossils from the Tertiary period confirm that they were once widespread over the whole of Eurasia and North America. The present Bering Strait was at that time dry land, and the climate then was such as to favor an exchange of fauna between Asia and North America. Fossils further show that the tapirs did not arrive in South America until the Pleistocene age. They are recent immigrants compared with such old inhabitants as the endemic mammals of South America (see p. 162).

The elephants which at present live in South Asia and Africa were still widely spread almost throughout the world in the Pleistocene age (Fig. 83). Their fossil remains are represented by many species from all continents except Australia and Antarctica. The same is true of the rhinoceroses which were found during the Tertiary period all over Africa, Eurasia, and North America, yet are presently restricted to a few species in relict locations (in Africa and Asia).

We know from numerous fossil finds that the original home of the camel types was the North American continent and that from there they reached Eurasia, Africa, and South America (llamas) in the course of the Pliocene-Pleistocene ages.

Araucarias, which today have a wide distribution as ornamental trees (*A. excelsa* — the Norfolk "fir") are natives of South America, East Australia, New Zealand, New Guinea, New Caledonia, and the Norfolk Islands; the southern beech occurs in Australia, Tasmania, New Zealand, New Caledonia, Patagonia, and South Chile. It is true that many fossil woods from Mesozoic deposits in the northern hemisphere have also been called *Araucarioxylon* because of their anatomical structure, but there is no proof that the trees belonged to the genus *Araucaria.* Fossil remains of *Nothofagus* are known from Australia, Zew Zealand, South America, and even from Antarctica. Thus, the southern beech, along with various fresh-water fish and land insects, is also regarded as providing support for the belief that there was once contact between South America, Antarctica, and Australia. Although it is not possible at present to pass definite judgement on the time at which the continental masses split asunder, there is no doubt at all that the Australian block (Australia and New Guinea) did not reach its present position until the Tertiary period.

The examples of disjunctive distribution in animals quoted here require, in order that the present pattern of distribution be achieved,

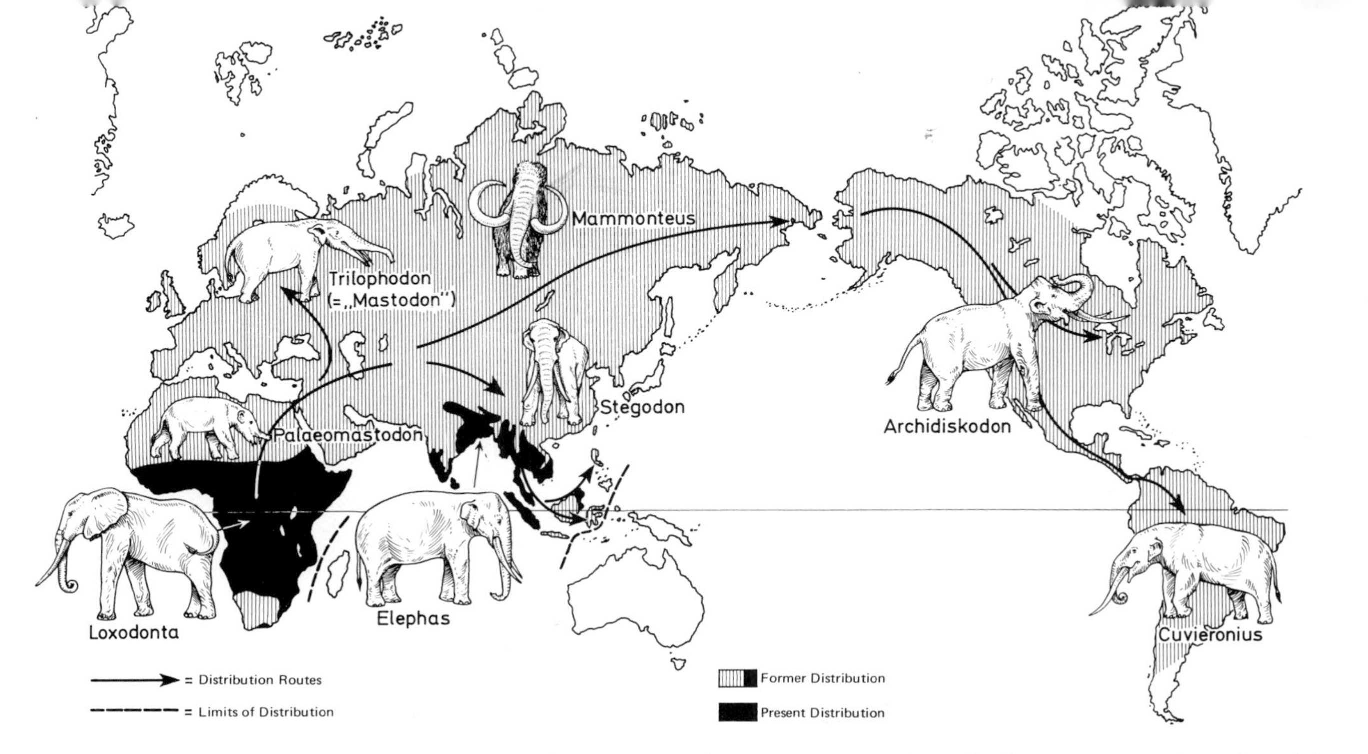

Fig. 83. Former and present distribution of Proboscideans: in the Late Tertiary natives of Africa, Eurasia, and North America; in the Pleistocene in South America as well. Disjunctive distribution in modern times (Africa: *Loxodonta africana;* South Asia: *Elephas maximus).* In the Late Tertiary and Quaternary spreading over the Arabian peninsula, Bering and Panama bridges.

that land bridges should have existed between continents. Such were and are the Panama bridge joining North and South America, the Bering Bridge between Asia and North America, and the Arabian peninsula joining Asia with Africa.

Ice Ages and Eustatic Fluctuations in Sea Level

Of the land bridges named above, only the isthmus of Panama and the Arabian peninsula exist today. Both are "filter bridges," in G. G. Simpson's sense. This name expresses the fact that such land bridges do not permit a free exchange of fauna but only a restricted one. Thus, at the present time the Panamanian isthmus is impassable

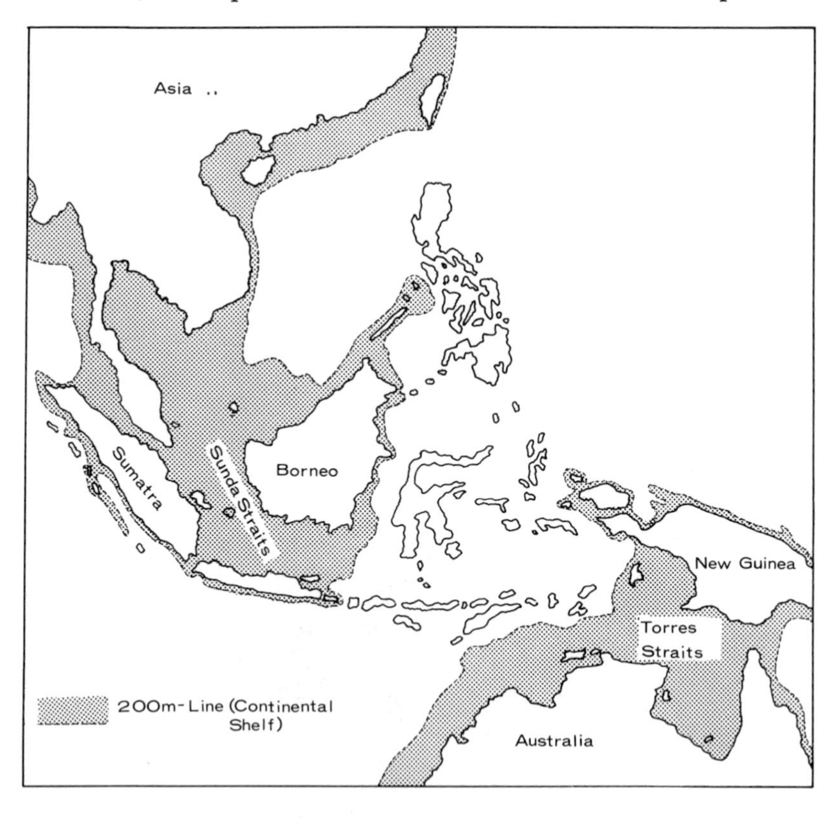

Fig. 84. Course of continental shelf in South East Asia and Australia (represented by 200 m depth line). Sunda Sea and Torres Straits partly landlocked in Ice Age.

for steppe dwellers, and the Arabian peninsula is accessible only to desert or semi-desert animals. Climatic conditions must have been different in the past, allowing the inhabitants of the open landscape to migrate from North to South America and forest dwellers from South Asia to Africa and vice versa.

During the Pleistocene age there were several alternations of warm and cold periods which were expressed in the subtropics as dry and rainy periods. Such changes of climate make it understandable that the "filter bridges" worked in different ways at different times; they also explain how former land bridges, like the Bering Bridge, could have originated.

During the cold periods there was a fall in the sea level due to the freezing of huge masses of water in the polar regions to form the mighty ice caps; the difference in sea level in cold and warm periods amounted to almost 200 m. Such a drop is sufficient to uncover large parts of the continental shelf covered by shallow seas — e. g., the North Sea, the Sunda Sea, and the Torres Straits — thus creating land bridges which linked continents (Bering Bridge) or which joined islands with the mainland — e. g., England, New Guinea, Tasmania, the Japanese archipelago, and Ceylon and Sunda islands (Fig. 84). Thus, many modern islands could have been settled by continental animals for whom the sea presented an impassable barrier. However, many larger animals were able to cross narrow straits, as witness the presence of elephants and hippopotami in dwarf forms from the Pleistocene age on various Mediterranean islands — Malta, Crete, Cyprus [3] — and Madagascar.

The glaciations present a particular problem of their own in paleogeography. Ice Ages are usually attested by a variety of inorganic climatic evidence, such as moraines, glacier-polished rocks, cryoturbation phenomena, erratic blocks and ice wedges, and by plants and animals, for instance, in the Pleistocene, Arctic plants; also musk ox, reindeer, willow grouse and Arctic fox. The best-known glaciations are that of the Pleistocene age, which is characterized by a number of alternations between cold and warm periods, and the Permo-Carboniferous glaciation, traces of which are to be found everywhere on what was then the southern continent. On the other hand, true glaciations

[3] The Mediterranean islands named were not linked to the mainland even during the cold periods of the Pleistocene. However, the fall in sea level mostly reduced them to narrow or shallow waters.

are not known to have occurred in either the Mesozoic or the Tertiary periods. Various hypotheses have been advanced to try to explain this phenomenon, based not only on terrestrial factors (relief hypothesis, ocean currents) but also on extra-terrestrial ones (changes in solar radiation due to dark clouds, etc.). Paleontology has no solution to offer to this problem. Nevertheless, there are many clues which suggest that the ice ages were dependent upon the polar continents and the corresponding sources of moisture in the form of the oceans, so that, here too, continental drift may offer a probable explanation. Doubtless, an additional influence was also exerted by ocean currents and the relative positions of the land masses. We have only to consider the Gulf Stream to see how closely such phenomena are linked with paleogeographical facts, in this case, the existence of the Panama Bridge, and the whole question of the opening of the North Atlantic ocean.

Fairly recent discoveries have shown that the North Atlantic was not always an ocean but that it began to assume its present form only in the Early Mesozoic era when North America and Europe were drifting apart. This realization is based upon the close correspondence between the nonmarine vertebrate fauna of both continents. It can only be explained if there had been land contact from the Devonian through to the Permian periods.

Fossil Land Vertebrates and Their Importance in Paleogeography

The present-day mammalian fauna of South America is characterized in its zoogeography by the large number of endemisms. South America is thus assessed — together with parts of Central America whose large animals have immigrated from South America since the Pleistocene — as a zoogeographic region (neotropis) on its own. The special position occupied by its fauna becomes comprehensible when we know the history of its animals which, again, reflects paleogeographic events. The history of the fauna is documented by fossil finds. From these it is clear that of the present fauna only the marsupials, the New-World monkey (Platyrrhina), the Xenarthra (armadillos, sloths, and anteaters), and the caviomorphs (including guinea pigs, maras, chinchillas, agutis, octodons, and tree porcupines) represent the old-established elements, while the even-toed animals so characteristic of South America (such as llamas, peccaries, and pudus), the odd-toed animals (tapirs), the predators (jaguars, wild cats, "foxes," bears, raccoons, and martens), leporids (wood rabbits), and probably also the hamster-like rodents (new-world mice) were late immigrants, mostly during the Pleistocene (Fig. 85).

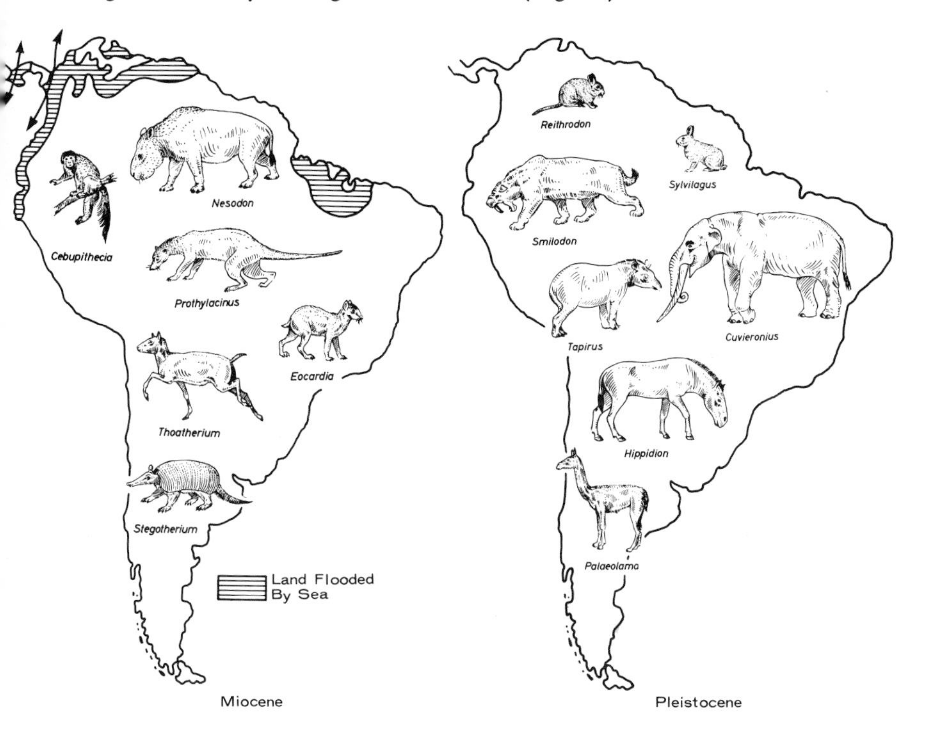

Fig. 85. Left: characteristic mammals of the Miocene of South America (marsupials: *Prothylacinus;* Xenarthra: *Stegotherium;* New-World monkeys: *Cebupithecia;* rodents: *Eocardia* and hoofed animals: *Nesodon)* as endemisms. Right: late immigrants in the Pleistocene (rodents: *Reithrodon;* leporids: *Sylvilagus;* sabre-toothed cats: *Smilodon;* tapirs: Tapirus; mastodons: Cuvieronius; horses: *Hippidion* and even-toed hoofed mammals: *Palaeolama).* Arrows show sea links.

The absence in South America of the groups of animals which were so characteristic of the Tertiary period in North America, their distribution as far as Panama, as attested by fossils, and the correspondence between the marine fauna of the Tertiary period in the East Pacific and the Caribbean — all these facts provide evidence enough that for almost the whole of the Tertiary period the continents of North and South America were separated by a narrow sea. The Panama Bridge did not appear until the very Late Tertiary period, thus faciliating an exchange of animals which is still going on today.

In the Pleistocene age, the animals just cited had already migrated into South America together with one-toed horses, mastodons, and sabre-toothed cats. There now appeared in Central and North America, as immigrants from the south, armadillos, marsupials, New-World monkeys, and caviomorph rodents.

Here then is evidence aplenty to show that the Panama Bridge has existed only since the Late Tertiary period, and hence that the Gulf Stream presumably also assumed its familiar course about this time. The Antilles appear, in the Tertiary period at any rate, to have been only occasionally important for floral and faunal distribution as a land bridge in the form of a chain of islands from Yucatan to northwestern South America.

The present-day distribution of monk seals (Mediterranean Sea–Caribbean Sea–Hawaiian Islands) and the earlier distribution of sea cows would be just as difficult to explain as the correspondence between the Tertiary marine fauna (e. g., pelecypods, snails, sea urchins, crabs, and coastal fishes) in the East Pacific and the Caribbean unless a Panama Strait had existed during the Tertiary period.

Fossil Marine Animals as Paleogeographic Evidence

The correspondence just mentioned between the East Pacific and Caribbean marine faunas is only one example of the importance of fossil marine fauna in matters of paleogeography. Even today, there are many marine forms with a distribution on both coasts of America, thus bearing witness to the former sea link across Central America. Another example is the Bering Straits, where the continual exchange of land mammals bears witness to the existence of a land bridge throughout almost the whole of the Tertiary period. It is not until the Pliocene epoch and the Quaternary period that Pacific marine forms are found on the north coast of Alaska, and these could not have reached this coast of the north polar sea unless the Bering Straits had been open.

Fossil marine fauna also enable us to draw conclusions about the former faunal provinces. Thus, the planktonic Foraminifera of the

Tertiary period allow actual climatic zones to be distinguished in the same way as the distribution of reef coral and the occurrence of nummulites permitted conclusions about the tropical area in those times. In the Mesozoic era it is in particular the ammonites, already familiar to us as index fossils, which enable the various climatic regions to be identified. Already in 1883, M. Neumayr had distinguished a boreal and a Mediterranean province in the Jurassic period on the basis of

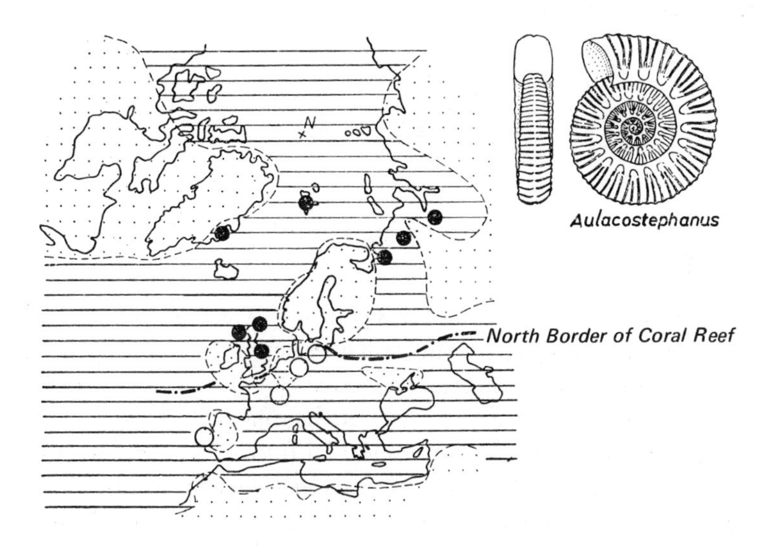

Fig. 86. Ammonites and paleogeography. Distribution of *Aulacostephanus (Xenostephanus)* as a boreal group and of the reef-building corals as tropical elements in the Upper Jurassic (Kimmeridgian) of Europe. Solid spots mark places where ammonites have been found; circles mark occurrence of reef-building corals whose species (over 110 in the south) decline towards only seven in the north. (After B. Ziegler, 1967, revised.)

ammonites, and now recent studies of ammonites and belemnites by W. J. Arkell. G. R. Stevens, and B. Ziegler have led to a much more extensive classification of the marine faunal provinces of that time (Fig. 86). They have further shown that ecological sources of error need to be excluded, since the various ammonites inhabit different depth zones and this could simulate climatically determined differences.

Such sources of error are virtually nonexistent for pseudo-planktonic organisms, i. e., those which drifted on seaweeds — particularly certain pelecypods *(Inoceramus, Halobia, Monotis)* — and used their byssal filaments to cling to plants floating near the surface.

For the later Paleozoic era a similar role in the differentiation of marine provinces may be assigned to the Fusulinids (Foraminifera), which often occur as rock-forming fossils.

These few examples must suffice to illustrate both the possibilities and the limitations of the assistance which fossils can provide in regard to paleogeography.

X. "Living Fossils"

What Are "Living Fossils"?

The expression "living fossils" was coined by Darwin who applied it to the East Asian ginkgo tree. The term does not mean fossils that have come back to life.[1] Rather it refers to present-day flora and fauna that are to be regarded as phylogenetic survivors, having evolved little or not at all in comparison with related primordial forms. The absolute geological age alone is, however, not conclusive for "living fossils," since the age of the lineage of such surviving forms may be anything from a few million to several hundred million years. What settles the matter is the small amount of phylogenetic alteration undergone by these forms since they originated, and the fact that they have often retained an ancient structure in comparison with other, faster evolving forms. However, it may happen — and this will become clear from the statement on p. 96 — that organisms retaining such phylogenetically primitive characteristics may at the same time possess other features which have been acquired later, some of them highly specialized. As a rule, the "living fossils" are rather isolated in the systematics of Recent forms, as their relatives have been extinct for a long time. Frequently, they are associated with a spatially limited distribution. Here, too, we shall offer some examples to provide practical illustrations.

---

[1] Some years ago living bacteria were reported in Paleozoic halite and, according to H. Dombrowski, were supposed to have remained preserved in the salt up to the present time.

Recent Coelacanths: Latimeria chalumnae

Certainly the most famous living fossil in the animal kingdom is the coelacanth *Latimeria chalumnae,* which was not found until 1938. It is the only living member of a geologically very old group of fishes. The discovery of this fish off the coast of South Africa was for several reasons a sensation in the world of science. On the basis of fossil finds and the fact that until 1938 there had been no sign of any Recent coelacanths, the crossopterygians were considered to be extinct. The

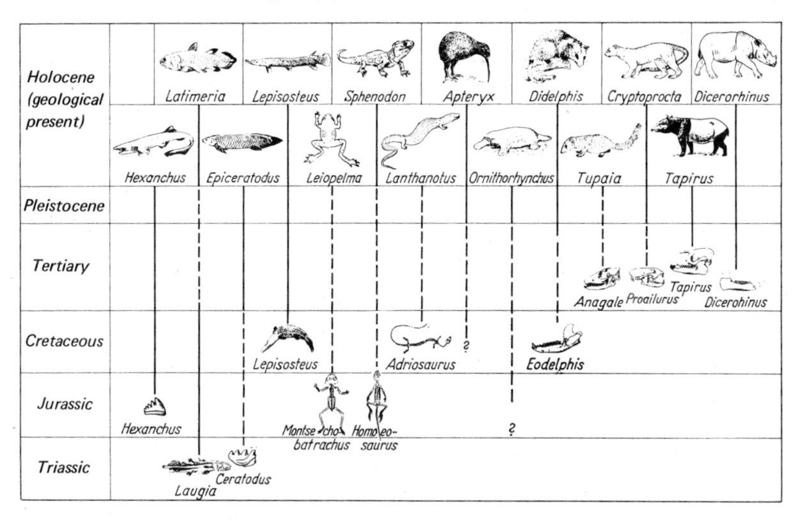

Fig. 87. The most important "living fossils" among the vertebrates linked with fossils of the geological past. Dotted line indicates fossil record missing.

geologically youngest fossils date from the Cretaceous period, and scholars consequently were in agreement that the coelacanths must have died out at the end of the Mesozoic era, hence more than 60 million years ago. But the evidence of a Recent coelacanth was very important for the life sciences for yet another reason. These fishes have certain characteristics which make it clear that they were the ancestral forms of all land vertebrates, hence, amphibians and reptiles and birds and mammals (see p. 109). In this respect, the zoologists were somewhat disappointed, as *Latimeria chalumnae* belongs to the evolutionarily unimportant lineage of coelacanths (hollow-spined) and not to the phylogenetically significant Rhipidistia, which had already disappeared by the end of the Paleozoic era. What greatly surprised the experts in addition was the fact that it differed so little from the fossil coelacanths of the Mesozoic era, not only in its general habitus, but also in the shape and position of its fins (Fig. 87). This phylogenetic conservatism is the most striking feature of living fossils, and it also provides information about the soft tissues of their ancestors, which have not been preserved in fossil form.

How is it that this living coelacanth, of which more than a dozen specimens have been caught since 1938, managed to escape the attention of scientists for so long? As it turns out, this coelacanth was well known to the natives of the Comores Archipelago to the north of Madagascar whose territory is the true habitat of these fishes. They use coelacanths for food, although they tend to confuse them with large wrasses. The natives also use the rough-surfaced scales of the fish as a kind of sandpaper to roughen the surface of bicycle tires before affixing a patch. Because these coelacanths live at depths of 150 to 800 m and are primarily seafloor dwellers, inhabiting the hollows and depressions of the steep submarine slopes of the islands, they remained hidden from the zoologists for so long. The first specimen to be caught had been carried far to the south by the current.

The migration of the coelacanth into such a habitat at the end of the Mesozoic era, at the latest, also explains why there were no fossils from the Tertiary period, despite the fact that Tertiary deposits are found all over the world and are certainly not poor in fossils.

In comparison with the Mesozoic coelacanths, the bone substance of *Latimeria chalumnae* shows considerable retrogression — the skeleton consists only of cartilage in some parts. A similar process is known in the lung fish, where phylogeny has produced an increasing substitution of cartilage for bone. The intestinal tract of *Latimeria chalumnae* is in the form of a spiral gut, which is characteristic of the cyclostomes, cartilaginous fish, and archaic types of bony fish (e. g., *Polypterus*, dipnoans). In contrast, the Paleozoic coelacanths have the lung which is present in addition to the gills converted into a kind of fat bag, as is also found in other deep-sea fish. *Latimeria chalumnae* breathes entirely through its gills, and it does not have a pharyngeal cavity so that the choanae — the openings from the nasal pharynx into the mouth cavity — are also missing. The choanae are found only in the Paleozoic rhipidistians, mentioned above, which, incidentally, were fresh-water fish. The coelacanths, to which *Latimeria chalumnae* belongs, were already salt-water inhabitants at the end of the Paleozoic era. For the reasons given, the anatomical structure of *Latimeria chalumnae* cannot be taken as a model for the reconstruction of those Paleozoic crossopterygians who were the ancestors of land vertebrates.

Lung Fish, Gar-Pikes, and Bowfins

Other living fossils among the fishes include the Australian lung fish and the North and Central American gar-pikes; all these are fresh-water fish, and only a few gar-pikes occasionally venture into salt water.

The Australian lung fish *Epiceratodus forsteri* (Fig. 87) is the oldest living form of lung fish (Dipnoi), and its habitus is scarcely differentiated from that of the lung fish of the Triassic period *(Ceratodus)*. It also has tasselled fins and a lung in addition to its gills, and because of this the lung fish were once regarded as the ancestors of land vertebrates. However, certain features in the structure of the cranium and dentition are now considered to make this improbable. The Australian lung fish, which, until it was transplanted into the lakes and reservoirs of Queensland, was found only in the Burnett and Mary Rivers of North East Australia, does not bury itself in a mud capsule like its African cousins described on p. 74.

Lung fish were distributed in the Paleozoic era in Europe and in North America as well. Thus, the occurrence of *Epiceratodus forsteri* is typically relict, like that of many other living fossils. The same is true of the gar-pikes *(Lepisosteus,* Fig. 87) of North and Central America, whose relatives covered the whole of the northern hemisphere in the Tertiary period. The genus *Lepisosteus* is known since the Cretaceous period. The gar-pikes, which have a stout armor consisting of rhomboid, nonoverlapping ganoid scales, belong to the ganoid fish;[2] in the Mesozoic era these fish played a role similar to that of the "modern" bony fish, the teleosts. The gar-pikes are some of the last representatives of this once very important group, the only survivors of which are the bowfin *Amia calva* and the sturgeons. The heavy scales of the gar-pikes are only one of the ancient characters which distinguish these fish from "modern" Teleostei.

---

[2] The scales of these fish are very strong and are coated with an enamel-like substance (ganoin), which is what gives them their name.

Neopilina galatheae, the "Primordial Mollusc"

Possibly an even greater sensation in zoological circles was the discovery in 1952 of *Neopilina galatheae*, the "primordial mollusc," dredged up in the Pacific Ocean off the coast of Costa Rica from a

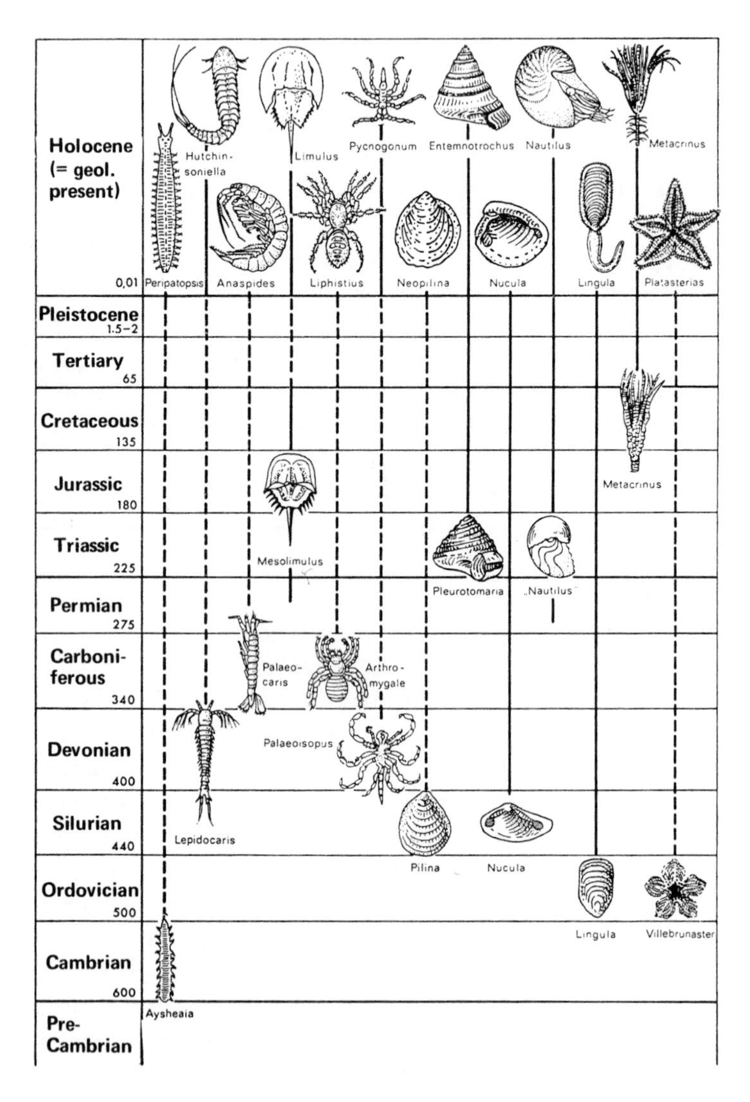

Fig. 88. Characteristic "living fossils" among the invertebrates and their fossil relatives. Dotted line indicates fossil record missing.

depth of 3,570 m. To the nonexpert such finds, which have since been duplicated at similar depths in other parts of the Pacific, may not seem particularly exciting (Fig. 88). These animals are invertebrates, molluscs with a noncoiled conical shell, rather like the limpets one sees everywhere along rocky coasts. As already mentioned on p. 109, the shells of *Neopilina galatheae* and its fossil relatives from the Paleozoic differ from those of true limpets by having five or six pairs of muscle scars instead of one horseshoe-shaped muscle scar. These correspond to divisions of the soft parts, which means that *Neopilina galatheae* and its fossil relatives are classified as a separate group within the invertebrates, the Monoplacophora. The importance of this discovery lies not only in the evidence it provides of a phylogenetically important group of forms thought to have been long extinct or whose continued existence was at best a matter for speculation. So far, fossil Monoplacophora have been confirmed only from the Early Paleozoic era, i. e., from an era over 350 million years back in time. This is a very primitive group of animals which, as finds to date give reason to suppose, are presently to be found only in the deep ocean, and which would more correctly be called "primordial conchifers" than "primordial molluscs." Conchifera include all molluscs with one- or two-piece shells — snails, pelecypods, scaphopods, and cephalopods.

Lingula and Crania

So far we have described the "primordial mollusc" *Neopilina* and the lung fish *Epiceratodus*, two living fossils whose lineages can be traced right back to the Paleozoic era, but whose living representatives either are not found as fossils or are found only back as far as the end of the Mesozoic era. We shall now mention a few living fossils which belong to genera known already from Paleozoic deposits. The best-known and longest lived genus is that of *Lingula:* these are brachiopods, a group of animals known to have existed right back to the Precambrian period and whose members, having a two-piece shell, look something like mussels (Fig. 88). *Lingula* and *Crania*, another genus of brachiopods, are found from the Ordovician period to the present and thus have, as genera, achieved a lifespan of some 430 million years. These are early types of brachiopod which do not have a hinge and whose pedicle protrudes from between the two halves of the shell. The "tongue mussels" — *Lingula anatina*, etc. — are widely distributed in the Indian and Pacific oceans and live in slime-covered vertical tubes on the seafloor beyond the low-water mark. The *Crania* species, on the other hand, have no pedicle but are firmly attached to the rocks with one shell and are found on rocky coasts of the northerly seas of Europe and western India.

Entomostraca as Living Fossils

The longest-lived Recent species is, however, the fresh-water crustacean *Triops cancriformis,* a branchiopod which is found as early as the Late Triassic period and is thought to have remained unchanged as a species ever since — that is, for 180 million years. These are the "leaf-foot" crustaceans (Phyllopoda = Branchiopoda). Zoologists have been much puzzled by the sudden appearance and disappearance of the branchiopods and by the fact that decades passed before a male was found. [3] They live in Central Europe and are found in fresh-water pools which dry out for some of the time. They do not usually appear until April, and they die when the pools dry up, assuring the continuation of the species by laying eggs in the hardening mud. After violent rainstorms, the crustaceans hatch in such large numbers that popular belief has it that they have fallen from the skies. The body of *Triops cancriformis* is covered by a shield-shaped shell up to 3 cm long and along its underside it has no fewer than 60 pairs of swimming feet; it swims on its back. The genus *Triops* itself dates from the Permian period, thus from the later Paleozoic era. However, according to their organization and their wide distribution, the Recent branchiopods cannot be designated living fossils. Only the dwarf crustaceans, some of which were discovered relatively recently (e. g., *Hutchinsoniella, Bathynella, Anaspides,* and *Nebalia)* may be regarded as living fossils, since their ancestors are known to have been alive in the Paleozoic era (Fig. 88).

Nautilus and Limulus

Phylogenetically conservative types which are also living fossils are found in the genera *Nautilus* (pearly Nautilus) and *Limulus* ("horseshoe crabs"), both ocean dwellers. A single species of Limulus occurs in brackish and fresh water in estuaries and rivers.

[3] Branchiopods can also reproduce by parthenogenesis, literally, virgin birth.

Nautilus is the only living genus of tetrabranchiate (four-gilled) cephalopods, which had a very rich variety of forms in the Paleozoic and Mesozoic eras and which, unlike the cuttlefish, have an external shell. The genus *Nautilus* itself is known only back to the Early Tertiary period (Eocene), but the type of the shell was already present in the Late Paleozoic era, as is clear from fossils. The genus is at present represented by very few species (Fig. 88), limited to the southwestern Pacific and inhabiting the seafloor in areas where the water is calm. A relict-type distribution, an isolated systematic position, and primitive characteristics mark the genus *Nautilus* as a true living fossil.

The same is true of *Limulus polyphemus*, the "horseshoe crab," which is not a crab at all but more closely related to scorpions and spiders. The king crabs, called "horseshoe crabs" in the United States from the shape of their dorsal shells, live on the Atlantic coast of North America. They are not found on the Pacific coast. Related forms called "molluccan crabs" are known from the Gulf of Bengal to the coasts of New Guinea, the Philippines, and Japan. The king crabs have a domed, two-piece dorsal shell, the front part, which is the larger, having the shape of a horseshoe in outline, whereas the rear part is almost triangular, with spines along its edge (Fig. 88). At the end of the body is the long, mobile, spiny tail. There are numerous pairs of limbs, some of which serve for locomotion, and others — through the gill leaves — as breathing organs or even as jaws, as they are equipped for chewing. The Recent king crabs are shallow-water forms which come into the vicinity of the coast to breed. King crabs or Xiphosura were widely distributed in the Paleozoic era and had a large number of genera; European forms are known to date from the Mesozoic era and the Teriatry period.

Sphenodon punctatus, the Tuatara

So far, we have dealt only with living fossils that inhabit the water. Since this could easily leave the reader with the erroneous idea that there are no living fossils on land, we shall select a few of the many terrestrial examples. Let us begin with *Sphenodon punctatus* (Fig. 87), a New Zealand lizard, called tuatara by the Maoris. So long as coelacanths and primordial molluscs remained undiscovered, the tuatara reigned undisputed as *the* "living fossil." Although the tuatara looks like a lizard, it is not one. It is the only living survivor of the group of Rhynchocephalia, widely distributed in the Mesozoic era, which have a skull shape that distinguishes them fundamentally from lizards because of its two temporal arches and temporal openings. The genus *Sphenodon* is descended from reptiles that were contemporary with the dinosaurs but that survived the end of the Mesozoic era, when the dinosaurs became extinct. *Sphenodon punctatus* is not very different from its Jurassic ancestors, and its skeleton retains the same general structure.

The tuatara, which is found today only on some of the small offshore islands off the coast of New Zealand, is the subject of a protection order. It lives with the procellariids in underground holes made mainly by the birds, partly in the soil and partly in the humus layer. The tuatara is most active at a temperature around 12° C. This seems astonishing for a reptile, but it has to be seen as an adaptive trait rather than as its original phylogenic behavior, since two native lizards and a gecko also live under the same temperature conditions.

Leiopelma and Apteryx

New Zealand harbors other living fossils: it is the home of the so-called primordial frog *Leiopelma*, and of the kiwi *Apteryx*, the flightless bird, which even the layman recognizes as a curiosity.

The frogs, represented by three species (Fig. 87), have primitive features that distinguish them anatomically from the other frog amphibians (structure of vertebrae, tail musculature, etc.) just as the kiwis are different from birds that are capable of flight. The nearest relatives of these frogs are found in the Jurassic period of Europe *(Montsechobatrachus)* and North America. They clearly had a wider distribution than is indicated by the fossils from that time. But even if we compared it with their former distribution on the basis of fossils, the present distribution of the primordial frogs is certainly a relict one.

The kiwis have, in addition to various primitive features, certain special characteristics which show a high degree of specialization (wing regression, feather structure, excellent sense of smell, etc.), and this specialization gives them a special systematic position. The nearest relatives of the kiwi *(Apteryx australis,* etc.) have been found as fossils on the Australian mainland.

Duckbill and Echidna

Australia is another continent that is the home of various living fossils and, of these, we have already mentioned the Australian lung fish. Then there are the duck-billed platypus and the echidna (Fig. 87), which are classified together by taxonomists as monotremes and differ from other mammals in their morphologic and physiologic peculiarities. Since fossils of monotremes are known only from Pleistocene deposits, the question is whether the monotremes really deserve to be called living fossils. The label is, however, justified by numerous primitive characteristics. Thus, the fact that there is only one body orifice for the discharge of the products of the kidneys and gut, as well as for reproductive products, is just as unique as the discovery that they lay eggs. They also have a number of anatomical peculiarities in the structure of the skull and shoulder girdle which they share with the reptiles, and these, too, must be regarded as primitive. These peculiarities have even caused it to be questioned whether the monotremes are truly mammals.

All these ancient features suggest that the phylum took a completely different path of development from that of the other mammals. Fossils of what are undoubtedly reptiles from the later Mesozoic era show that certain skeletal features have remained unchanged since that time. Many of the features characteristic of Recent monotremes are later adaptive phenomena — horny beak, loss of teeth and swimming feet in the duckbills *Ornithorhynchus anatinus;* spines, feeding habits, and an elongated muzzle in the echidnas *Zaglossus* and *Tachyglossus* — so that it has not been possible to use either of these creatures as models for the reconstruction of Mesozoic monotremes. This is something which has been found with other living fossils.

Peripatus and Lanthanotus

Other living fossils of less attractive appearance are the worm-like onychophorans, the "earless" monitor lizard and the Liphistiidae (primitive spiders). The onychophorans, or claw-bearers, look from the outside like caterpillars or segmented worms, are only a few centimeters long and live in fallen leaves or rotting wood, or under stones. So far they have been found in every continent except Europe. Many are restricted to the southern hemisphere — South Africa, Chile,

Australia, and New Zealand. They are very primitive animals which combine the features of segmented worms (annelids) with those of arthropods (insects, spiders, crabs, etc); they must be looked upon as the last descendents of a very ancient lineage. Their nearest relatives, known from the very old marine deposits of the Cambrian period, have been described as *Aysheaia* and are very little different from the Recent *Peripatus* (Fig. 88), the best-known living genus. Although the Recent species are land animals, they are active only when air humidity is high. A similarly concealed existence is lived by the "earless" monitor lizard *Lanthanotus borneensis* (Fig. 87) of Sarawak, a lizard not considered worthy of a second glance by the uninformed layman. From its external appearance, only its scales and the minuteness of its legs, which are hardly used at all for locomotion, are likely to strike the observer as unusual. The creature moves by undulating its very long body; this was discovered fairly recently when living specimens were caught. However, the really interesting thing about *Lanthanotus* is its anatomical structure which shows features related to those of the Cretaceous reptiles Dolichosauridae and Aigialosauridae. At the same time, *Lanthanotus borneensis* displays features which are characteristic of certain snakes first found as fossils in the Cretaceous period, e.g., giant snakes. Of course, the "earless" monitor lizard cannot be regarded as the ancestor of the snakes, but it is very closely related to the order. Its present occurrence in Borneo is of the relict type and is remarkable in comparison with that of its Cretaceous ancestors; this, together with its curious anatomy, stamp it as a true living fossil.

Among Recent spiders the Liphistiidae represent some of the most ancient forms. They are limited to a few species in South East Asia and are distinguished from other living spiders by their articulated hind parts and the number and position of their spinning glands, in which, however, they resemble the spiders of the Carboniferous period (Arthrolycosidae).

Opossum and Sumatran Rhinoceros

There are also some living fossils among the mammals which deserve a mention, such as the American opossum, which represents the mammals of the Cretaceous period, or the Asiatic two-horned rhinos of South Eastern Asia. To judge from fossils, present-day opossums differ very little from their ancestors in the Cretaceous period. Among the rhinoceros, the genus *Dicerorhinus* has species going back to the Early Tertiary period (Oligocene) which are hardly different from the extant Sumatran rhinoceros *Dicerorhinus sumatrensis* (Fig. 87). This is the living fossil among Recent rhinos.

We shall conclude our examples of animal living fossils with the Sumatran rhinoceros, although this by no means signifies that there are no other animals worthy of this description — the list could be extended indefinitely.

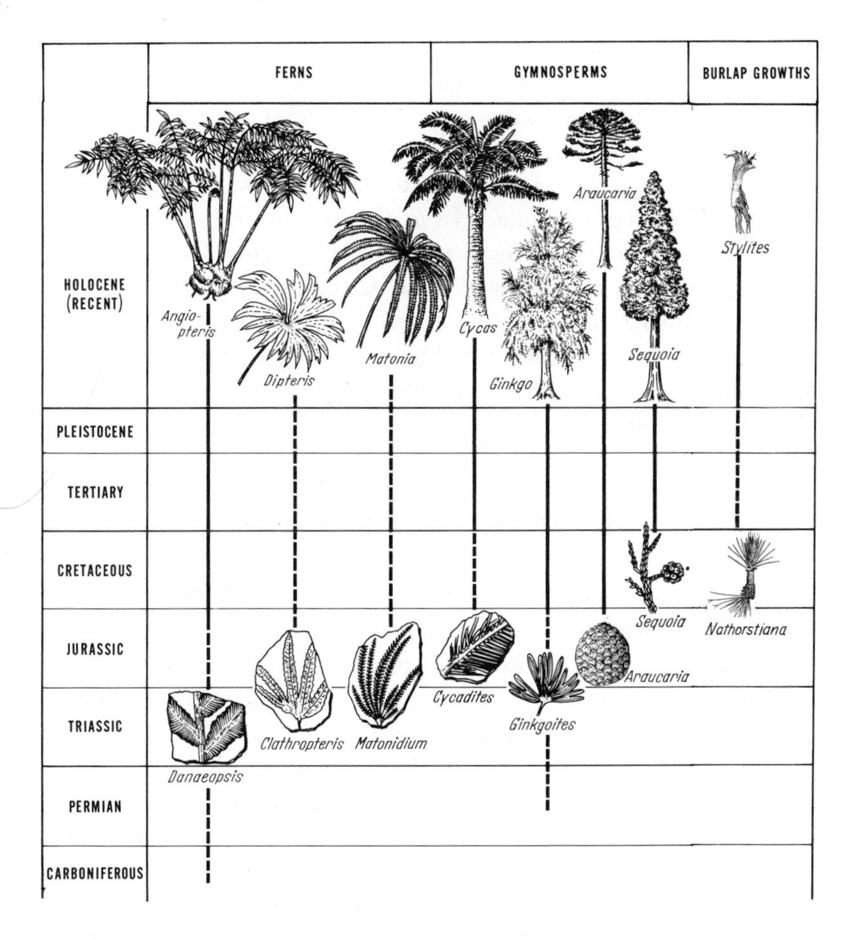

Fig. 89. "Living fossils" among plants and their fossil relatives. *Angiopteris* is a representative of the Marattiaceae and *Stylites* is a member of the Isoetaceae (quillworts).

Ginkgo biloba, Cycads, Sequoia, Araucaria, and Ferns

The best-known living fossil of the plant kingdom is certainly the ginkgo tree, *Ginkgo biloba.* This was discovered in historical times as a relict in East Asia and is now grown as a park and ornamental tree in all parts of the world. Although its leaves give it the look of a deciduous tree, it is in fact a gymnosperm. Its method of fertilization is not by means of pollen tubes, like that of the conifers, but by motile spermatozoids, evidence of rather primitive behavior. The ginkgoales were widespread in the Mesozoic era, and *Ginkgo biloba* (Fig. 89) is the only surviving member of this group once so abundant in species. In its wild form, it was definitely limited to a relict area.

The history of the "palm ferns" or Cycadeae (e. g., *Cycas revoluta,* the false sago palm), which also enjoyed a wide distribution with a large number of species in the Mesozoic era, is similar. The big trees *(Metasequoia glyptostroboides* from China and *Sequoiadendron giganteum* and the redwood *Sequoia sempervirens* from California) and the araucarias are often looked upon as the living fossils among the conifers. Among the ferns there are both the primitive tree ferns Cyatheaceae and Marattiaceae (Fig. 89) and the Dipteridaceae and Matoniaceae, which are now limited to a few species in the Indo-Malaysian tropics.

How Do "Living Fossils" Happen?

What are the possible reasons for the survival of "living fossils"? What they all have in common is their phylogenic conservatism over long periods of time. Is this due to external influence or are there inner reasons, or do both play some part? It is not at all easy for the paleontologist to answer such questions, but let us try. Living fossils are found within all the larger phylogenetic units. They are found in the most varied habitats, from deep-sea biotopes to coastal regions, from ground water to rivers and lakes, in tropical jungles just as much as in open landscapes. Although a habitat where living conditions have remained constant is a necessary condition for the existence of phylogenetically conservative types, this alone is not the decisive factor. The argument that it is always simply individual cases, but never whole faunas and floras which survive through millions of years also lends support to the view that other factors are involved. This again is clear from the spatial distribution of living fossils. Thus, New Zealand is primarily *the* island where such forms are found, and this leads to the assumption that isolation has something to do with the matter, if only by excluding competitors. The counterpart to island isolation in terrestrial species is the case of limnic forms in lakes. Hence, the faunas of Lake Baikal in Central Asia, of Lake Ochrid in South East Europe, and of Lake Nyasa in East Africa are basically of the relict type and are distinguished by a large number of endemic species.

This whole complex of questions is related to another which has already been briefly touched upon in another context, that of extinction. The forms which prove to be the longest lived are usually not particularly specialized. Furthermore, they occur in habitats that are not very liable to be exposed to sudden change, such as the depths of the sea as against the surface, ground water, and the mesopsammon, and the humus layer in primordial forests as against steppes and deserts, and so on. Even the habitat of the branchiopods (pools which dry out), which from a certain point of view might be termed an extreme biotope, is conservative to the extent that it excludes predators (e. g., fish) or larger competitors for food.

To sum up, we may state on the basis of all this evidence that a variety of factors are vitally involved in the conservation and non-extinction of phylogenetically conservative types and that a *single* cause is never responsible. Their persistence until the present is of the greatest interest not only for zoologists and botanists concerned with problems of evolution, but also for paleontologists. For these "living fossils" give valuable insights into the structure of the soft tissues, which are not retained in fossil form.

Thus, "living fossils" constitute a point of contact which is just as valuable in teaching us about the plants and animals of the geological past as the remains of these organisms, that is, the fossils themselves.

It is hoped that this book has awakened an interest and some understanding of paleontology among a wider audience and has demonstrated that, although fossils are indeed dead, they may be called back to life by informed observation to yield a wealth of revelations, and that paleontology, far from being a dry museum science, possesses great practical importance.

Geological Time Scale

Geological Era	Period	Epoch or Age	Millions of Years Ago	Evolution of animals and plants	Era
CENOZOIC	QUATERNARY	Holocene (Recent) Pleistocene (Ice Age)	2—2.5	AGE OF MAMMALS Evolution of man	CENOPHYTIC
	TERTIARY	Pliocene Miocene Oligocene Eocene Paleocene	65	AND ANGIOSPERMS Evolution of mammals	
MESOZOIC	CRETACEOUS	Late Early	135	Extinction of Dinosaurs, flying reptiles, ichthyosaurs and ammonites	
	JURASSIC	Malm Dogger Lias	195	AGE OF *Archaeopteryx* (primeval bird) REPTILES	MESOPHYTIC
	TRIASSIC	Rhaetian Norian Karnian Ladinian Anisian Skythian	225	First mammals AND GYMNOSPERMS	

Era	Period		Epoch	Million years	Life	Plant age
PALEOZOIC	PERMIAN		Ochoan Guadalupian Leonardian Wolfcampanian	270	Extinction of Trilobites First mammal-like reptiles	PTERIDOPHYTIC
	CARBON-IFEROUS	PENNSYL-VANIAN	Virgilian Missourian Desmoinesian Atokan Morrowan		Carboniferous forests (ferns, club mosses and horsetails)	
		MISSIS-SIPPIAN	Chesterian Meramecian Osagean	340	First reptiles	
	DEVONIAN			390	First amphibians, earliest insects, Psilophytes	ALGOPHYTIC
	SILURIAN			430	First land plants	
	ORDOVICIAN			480	First fish (jawless)	
	CAMBRIAN			570	Invertebrates and marine plants	
	PRECAMBRIAN			5000	Earliest fossils	

Bibliography *

ABEL, O.: Lebensbilder aus der Tierwelt der Vorzeit, 2nd ed. Jena: G. Fischer 1927.

——— Vorzeitliche Lebensspuren. Jena: G. Fischer 1935.

——— Vorzeitliche Tierreste im deutschen Mythus, Brauchtum und Volksglauben. Jena: G. Fischer 1939.

AGER, D. V.: Principles of Paleoecology. New York: McGraw-Hill Book Co., Inc. 1963.

BOWEN, R.: Paleotemperature Analysis. Methods in Geochemistry and Geophysics, 2nd ed. Amsterdam: Elsevier Publ. Co. 1966.

BRINKMANN, R.: Abriss der Geologie II. Historische Geologie, 8th ed. Stuttgart: Enke 1959.

BROUWER, A.: General Palaeontology. Edinburgh and London: Oliver & Boyd 1967.

BURTON, M.: Living fossils. The Past in the Present. London and New York: Thames & Hudson 1954.

COLBERT, E. H.: Evolution of the Vertebrates. A history of the backboned animals through time, 2nd ed. New York: J. Wiley & Sons, Inc. 1969.

DARRAH, W. C.: Principles of Palaeobotany, 2nd ed. New York: Ronald Press 1960.

DELEVORYAS, TH.: Morphology and Evolution of Fossil Plants. New York: Holt, Rinehart & Winston 1963.

EASTON, W. H.: Invertebrate Paleontology. New York: Harper & Brothers, Publ. 1960.

GLAESSNER, M. F.: Principles of Micropaleontology. New York: J. Wiley & Sons, Inc. 1948.

GOTHAN, W., and H. WEYLAND: Lehrbuch der Paläobotanik, 2nd ed. Berlin: Akademie-Verlag 1964.

HEBERER, G. (ed.): Die Evolution der Organismen I, 3rd ed. Stuttgart: G. Fischer 1967.

HECKER, R. F.: Bases de la Paléoécologie. Paris: Edit. Technip. 160.

* See also the books and journal articles in the list of sources of the illustrations.

Hurley, P. M.: "The Confirmation of Continental Drift," *Scient Amer.* **4**: 53–64 (1968).

Irving, E.: Palaeomagnetism and its Application to Geological and Geophysical Problems. New York: J. Wiley & Sons, Inc. 1964.

Kay, M., and E. H. Colbert: Stratigraphy and Life History. New York: J. Wiley & Sons, Inc. 1965.

Koenigswald, G. H. R. von: Die Geschichte des Menschen, 2nd ed. Berlin: Springer-Verlag Inc. 1968.

Mägdefrau, K.: Paläobiologie der Pflanzen, 4th ed. Jena: G. Fischer 1968.

Martin, P. S., and H. W. Wright (eds.): Pleistocene Extinctions. The Search for a Cause. New Haven: Yale Univ. Press 1967.

Mayr, E., E. G. Linsley, and R. L. Usinger: Methods and Principles of Systematic Zoology. New York: McGraw-Hill Book Co. 1953.

Moore, R. C. (ed.): Treatise on Invertebrate Paleontology. Lawrence: Kansas Univ. Press 1953.

———, C. G. Lalicker, and A. G. Fisher: Invertebrate Fossils. New York: McGraw-Hill Book Co. 1952.

Müller, A. H.: Lehrbuch der Paläozoologie I–III. Jena: G. Fischer 1957 to 1970.

Nairn, A. E. M. (ed.): Problems in Palaeoclimatology. New York: Interscience Wiley 1965.

Neaverson, E.: Stratigraphical Paleontology. A Study of Ancient Life-Provinces, 2nd ed. Oxford: Clarendon Press 1955.

Piveteau, J. (ed.): Traité de Paléontologie I–VII. Paris: Masson et Cie. 1952–1970.

Pokorny, V.: Principles of Zoological Micropaleontology I–II. Oxford: Clarendon Press 1962–1965.

Romer, A. S.: Vertebrate Paleontology, 3rd ed. Chicago: Chicago Univ. Press 1966.

Runcorn, S. K. (ed.): "Continental Drift." *Internat. Geophys. Ser.* **3** New York (1962).

Schäfer, W.: Aktuo-Paläontologie nach Studien in der Nordsee. Frankfurt: W. Kramer 1962.

Schindewolf, O. H.: Grundfragen der Paläontologie. Stuttgart: Schweizerbart 1950.

Schuchert, Ch.: Atlas of Paleogeographic Maps of North America. New York: J. Wiley & Sons, Inc. 1955.

Schwarzbach, M.: Das Klima der Vorzeit. Eine Einführung in die Paläoklimatologie, 2nd ed. Stuttgart: Enke 1961.

Simpson, G. G.: Tempo and Mode in Evolution. New York: Columbia Univ. Press 1944.

SIMPSON, G. G.: Life of the Past. An introduction to Paleontology. New Haven: Yale Univ. Press 1953.

——— The Major Features of Evolution. New York: Columbia Univ. Press 1953.

STOCK, CH.: Rancho La Brea. A Record of Pleistocene Life in California, *Los Angeles County Mus. Sci. Ser.* 13, Paleont. **8**, Los Angeles (1949).

SYMPOSIUM: Gondwanaland revisited: New Evidence for Continental Drift, *Proc. Amer. Philos. Soc.* 112, **5**, Philadelphia (1968).

THENIUS, E.: Lebende Fossilien. Zeugen vergangener Welten. Stuttgart: Franckh'sche Verlagshandlung 1965.

——— Phylogenie der Mammalia (Stammgeschichte der Säugetiere einschliesslich der Hominiden). Berlin: W. de Gruyter 1969.

ZIMMERMANN, W.: Geschichte der Pflanzen, 2nd ed. Stuttgart: G. Thieme 1969.

List of Sources of Illustrations
(other than original photographs or drawings)

Figs. 4, 17, 18 — BACHMAYER, F., Universum, Natur und Technik, Sonderheft 1957, Wien.

Fig. 6 a—b — KIESLINGER, A., Natur und Technik 1, Wien 1947.

Fig. 6 c — AMMON, L. v., Abh., Bayer. Akad. Wiss., math.-naturw. Kl. 15, München 1886.

Fig. 8 — VOIGT, E., Nova Acta Leop., n. F. 3, No 14, Halle 1935.

Fig. 9 — VOIGT, E., Nova Acta Leop., n. F. 6, no 34, Halle 1938.

Fig. 10 — PFIZENMAYER, C. W., Mammutleichen und Urwaldmenschen in NO-Sibirien. — Leipzig 1926.

Fig. 13 — ABEL, O., Vorzeitliche Lebensspuren. — Jena 1935.

Figs. 22, 23, 78 — ZAPFE, H., Natur und Volk 87, Frankfurt 1957.

Fig. 25 c — BACHMAYER, F., Natur und Technik 13, Wien 1958.

Fig. 26 — ACCORDI, B., und R. COLACICCHI, Geol. Romana 1. — Rome 1962.

Fig. 27 b — ABEL, O., Vorzeitliche Tierreste im Deutschen Mythos, Brauchtum und Volksglauben. — Jena 1939.

Fig. 30 — KREJCI-GRAF, K., Erdöl. — Berlin-Göttingen-Heidelberg: Springer 1955.

Figs. 31, 43 b — PETRASCHECK, W. E., Kohle. — Berlin-Göttingen-Heidelberg: Springer 1956.

Fig. 32 — LEHMANN, R., Eclogae geol. Helv. 54, Basel 1961.

Fig. 33 a — BLACK, M., and B. BARNES, J. Roy. Microscop. Soc. 80, 1961.

Fig. 33 b — BLACK, M., Geol. Magaz. 99, London 1962.

Fig. 34 a—b — BACHMANN, A., und A. KECK, Mikrokosmos. — Stuttgart 1969.

Fig. 35 a—b — LEHMANN, W. M., Jber u. Mitt. o-rhein. Geol. Ver., n. F. 27, Stuttgart 1938.

Fig. 35 c — STÜRMER, W., Paläont. Z. 43, Stuttgart 1969.

Fig. 36 a — ROMER, A. S., Vertebrate Paleontology. — Chicago 1953.

Fig. 36 b — KUHN-SCHNYDER, E., Die Geschichte der Wirbeltiere. — Basel 1953.

Fig. 37 — MÄGDEFRAU, K., Paläobiologie der Pflanzen. — Jena 1956.

Fig. 39. — Gross, W., Paläont. Z 31, Stuttgart 1957.
Fig. 40 — Osborn, H. F., The Titanotheres. — Washington 1929.
Fig. 41 — Abel, O., Lebensbilder aus der Tierwelt der Vorzeit. — Jena 1927.
Fig. 43 a—c — Kräusel, R., Versunkene Floren. — Frankfurt 1950.
Figs. 44, 80 — Augusta, J., u. Z. Burian, Tiere der Urzeit. — Prag 1956.
Figs. 46, 59 — Koenigswald, G. H. R. v., Die Geschichte des Menschen. — Berlin-New York-Heidelberg: Springer 1968.
Fig. 47 a — Simons, E. L., Postilla 57, New Haven 1961.
Fig. 47 b — Simons, E. L., Proc. Nation. Acad. Sci. 51, Philadelphia 1964.
Fig. 48 — Thenius, E., Niederösterreich im Wandel der Zeiten. — Wien 1962.
Figs. 50, 51, 53, 82 — Thenius, E., Paläontologie. Die Geschichte unserer Tier- und Pflanzenwelt. — Stuttgart 1970.
Fig. 52 — Charrier, O., Atti Rassegn. Tecn. Soc. Ingegn. and Archit. Torino, n. s. A 19. — Torino 1965. Hopping, C. A.: Rev. Palaeobot. and Palyn. 2. — Amsterdam 1967.
Figs. 54, 55 — Grabert, B., Abh. Senckenberg. naturf. Ges. 496, Frankfurt 1957.
Fig. 56 — Moore, R. C., Treatise on Invertebrate Paleontology, Part L, Mollusca 4: Ammonoidea. — Kansas 1957.
Fig. 58 — Simpson, G. G., Horses. — New York 1951.
Fig. 60. — Thenius, E., u. H. Hofer, Stammesgeschichte der Säugetiere. — Berlin-Göttingen-Heidelberg: Springer 1960.
Fig. 62 — Thenius, E.: Lebende Fossilien. — Stuttgart 1965.
Fig. 63 — Wiedmann, J., Biol. Review 44. — Cambridge 1969.
Fig. 64 — Jarvik, E., Théories de l'évolution des vertébrés. — Paris 1960.
Fig. 66 — Glaessner, M. F., and B. Daily, Rec. S-Austral. Mus. 13, Adelaide 1959.
Figs. 68, 69 a — Soergel, W., Die Fährten der Chirotheria. — Jena 1925.
Fig. 69 c—d — Huene, F. v., Die fossilen Reptilien des südamerikanischen Gondwanalandes. — München 1942.
Fig. 71 — Zapfe, H., Palaeobiologica 7, Wien 1939.
Fig. 72 — Thenius, E., Carinthia II/151, Klagenfurt 1961.
Fig. 73 — Zapfe, H., Sitz.-Ber. österr. Akad. Wiss., math.-naturw. Kl. I, 155, Wien 1947.
Fig. 74 b — Slijper, E. J., Riesen des Meeres. — Berlin-Göttingen-Heidelberg: Springer 1962.
Fig. 76 — Seilacher, A., Marine Geol. 5. — Amsterdam 1967.
Fig. 81 — Hallam, A., Palaeogeogr., Palaeoclimat. and Palaeoecol. 3. — Amsterdam 1967.
Fig. 86 — Ziegler, B., Geol. Rundschau 56. — Stuttgart 1967.

Name and Subject Index

* See Bibliography.

† Numbers in bold face refer to illustrations.